The Fire Dogs of Climate Change

An Inspirational Call to Action

The Fire Dogs of Climate Change

An Inspirational Call to Action

Sally Andrew

FINDHORN PRESS

© Sally Andrew 2009

http://firedogs.findhornpress.com

The right of Sally Andrew to be identified as the author of this work
has been asserted by her in accordance with the Copyright,
Designs and Patents Act 1998.

Published in 2009 by Findhorn Press, Scotland

Pages 49, 72, & 86 * "Excerpted from the speech delivered by
Al Gore on July 17, 2008 on Renewable Energy at
The Daughters of the American Revolution's Constitution Hall.
Copyright © 2008 by Al Gore, reprinted with permission of
The Wylie Agency, Inc."

ISBN 978-1-84409-162-1

All rights reserved.
The contents of this book may not be reproduced in any form,
except for short extracts for quotation or review, without the written
permission of the publisher.

Edited by Michael Hawkins
Cover and interior design by Damian Keenan
Printed and bound in the European Union

1 2 3 4 5 6 7 8 9 10 11 12 14 13 12 11 10 09

Published by
Findhorn Press
305a The Park, Findhorn
Forres IV36 3TE
Scotland, UK

Telephone
+44 (0)1309 690 582
Fax
+44 (0)131 777 2711

info@findhornpress.com
www.findhornpress.com

About the Author

Sally Andrew has been involved since the 1980s in activism, adult education and writing for community organizations. She has a Masters in Adult Education (UCT) and a certificate in Environmental Education (Rhodes). This is her first book of creative non-fiction. She is currently writing a novel. She lives in the seaside suburb of Muizenberg, Cape Town with her man, cat and chameleons.

This book is dedicated to my dog, Riska.

Contents

Introduction 1
Foreword by Gaia 3

Reading Stories to Fire Dogs 5

Tails of Integrity 11

Fact Sheet 1: The Problem –
Environmental Destruction and Climate Change 18

Riding the Moon 23

Fact Sheet 2: The Solutions –
Energy Efficiency and Renewable Energy 30

Running at the Enemy 42

Inspiring Examples: Action 47

Dancing with Fire 66

Inspiring Examples: Groundbreaking Technology 71
Inspiring Examples: Sustainable Living 98

How Can <u>You</u> Make a Difference? 110
Climate Change Petition 112

Afterword by a Fire Goat 114

Educator's Guide 116
Thanks 122
References and Resources 124
Endnotes 130

The Fire Dogs of Climate Change

PEFC - 16-33-250

This book is printed on paper that
includes a minimum of 70 % of PEFC
certified material from forest which
has been certified against a PEFC
endorsed forest certification scheme
as sustainably managed.
www.pefc.org

Introduction

In the Chinese Year of the Fire Dog (February 2006–2007), I opened my heart a little to the Earth. I was struck by the heavy realization that life on this planet was in danger – as a result of human actions. *The Fire Dogs of Climate Change* is a result of my process of facing up to the environmental crisis – climate change in particular.

The stories in this book capture my emotional journey during this Fire Dog year: a journey in which I travelled through despair and powerlessness, and found places where I could be and act with strength and joy, in support of the Earth.

The fact sheets arise from my intellectual search for the information that we need to understand the problems and the solutions to climate change.

Then, in recent months, I travelled around the world (largely in the comfort of my own chair) searching for practical ways of tackling climate change. I googled and goggled across cyberspace, and emailed thousands of climate change professionals and activists. I collected hundreds of inspiring examples of political action, groundbreaking technology and sustainable living projects. I sifted through these and have laid out a few glowing embers for you to see.

I've also included an educators' guide for those who are able to spread this knowledge further.

I hope that the tools and tales in *The Fire Dogs of Climate Change* reach some part of you: your heart, mind, body or spirit; and inspire you to find your own way of protecting the Earth.

Foreword by Gaia

My name is Gaia. I have a kiss as gentle as butterfly eyelashes and a bite as deep as the jaws of a lioness in the soft neck of a young impala.

When you feel your feet each time they touch the earth, and you walk across plains and mountains until sweat pours down your limbs, making small rivers run through the dust on your body, and your eye takes in the millions of singing blades of grass, and your fingers stroke them gently, changing the way that the light paints each soft blade; then you are tracing the surface of my body, touching lightly the patterns of my hair.

When you throw yourself into the wildest of waves in a stormy sea, when you let them crash against your body, and you breathe the spray deep into your lungs, you are tasting the salt of my body.

I am in the impala when it grazes, when it strokes the air with its nostrils, and when it gives birth. I am its leap through the air as it kicks its legs back, flying across the grassy plains. I am the thorn trees in new leaf, between which the leaping buck moves. I am the hungry lioness stalking the impala. I am her slow stealthy movement, the sharp focus in her eyes. I am the frozen moment before she leaps and chases. I am the blood pumping in the impala as it runs, and the claws of the lioness as they sink into the hard muscles of the impala's flank. I am the fall and the struggle, the powerful weight of the lioness. I am the cry of warning that the buck calls out to its herd. I am the grip of the lionesses' jaw as it closes around the underside of the impala's throat. I am the soft neck of the buck that, in the final moment, offers itself up.

I am the love they both feel.

I am the heavy rain and the wild wind of the storm that pummels the earth, smashing trees and windows. I am the giant racing waves that rise without warning and drown steel and soil and flesh. I am the thousand ripping arms of the hurricane that tears across water and land. This is my breath.

I am the impala, and my flanks have been raked deep with claws. I am the hungry lioness and I can kill you in a moment. But I don't.

I offer you my throat.

Your name is Gaia, too.

The Fire Dogs of Climate Change

Reading Stories
to Fire Dogs

I have never been quite sure what to do when I grow up. When I was a little girl I played at being a fire engine, a nurse and a prime minister. I knew I had to do Something. I never would have guessed that my job would be to read stories to dogs.

When I was eight I did Schoolgirl Things. During break-time, my friend and I would go to a little wooden shack that smelled of earth and boys' socks.

We sat on the floor, on either side of our boyfriend, taking turns to kiss him. In between kissing we'd smoke cigarettes. Our innocent tongues were not tasting sweet or stale, but only the daring flavour of the unfamiliar.

I had stolen the cigarettes, Courtleigh Lights, from my mother. They fitted very neatly into the slots in my pencil-crayon tin. I thought I was doing a good job of making sure she didn't notice, by carefully taking only one or two from each of her different stashes. But obviously she had nothing better to do than count her cigarettes and soon announced that someone was stealing them.

When none of us owned up, she mumbled: 'That Elsie...'

This was the name of the domestic worker she employed. Elsie would bring us hot *rooibos*[1] tea in the mornings and clean my room – properly – not just by spreading my leopard-spotted blanket on top of the mess, as I would do.

'I'm going to have to fire Elsie,' my mother muttered. 'I can't have her stealing my things.'

I could sense a great injustice coming, and it was all my fault. I had to do Something!

'Mum, Mum, I know who's been taking the cigarettes. Come and look here,' I said.

There in his wicker basket was Gideon, our big Airedale puppy, with chewed up cigarettes strewn around him.

'Look, here's one of your cigarette boxes with Gideon's teethmarks in it.'

And indeed there were his teethmarks. Gideon looked up at us with his soft brown eyes, and his imploring woolly eyebrows. On his sweet puppy-breath was the tang of something sharp, and hanging off his lower gums was a strand of tobacco-drool.

5

'He's just a puppy, Mum. He chews everything. You've got to keep things out of his reach.'

No more cigarettes went missing. And nobody was fired, or spanked.

Some time later my mother told me she had known I was stealing the cigarettes, and the threat to fire Elsie was a ploy. She knew that I would do Something.

Later that same year I learned from my mother that police were shooting school-children who were marching in the streets. The year was 1976, and it was time to do Something. At my school we didn't march – we practiced hiding from the marchers. We had lots of 'riot drills': a very loud alarm bell would ring and we'd all rush into lines and march single file into the basement. Or the teacher would blow a little whistle, pprrrt! – and we would quickly pull our desks onto their sides and lie flat and quietly next to them. Listening to the sound of our own heartbeats.

One morning my mother gave me a black ribbon to tie around my arm. It was a mourning band for the children who were being shot dead.

I was a small girl with a single black ribbon around my skinny arm. But I managed to scare all the big boys at the school. One boy was particularly large and pink. He travelled in a pack.

'That's for Black Power isn't it?' he said. 'You want the blacks to come and kill us all?!'

He told me that the vice-principal wanted to see me. When I arrived in Mr Lindenboom's room, it was full of older students. They made a narrow pathway for me to walk to his desk. They stood all around and above me, like a silent ig-loo. My little body was hot and my face was red. Above the sound of the pound-ing in my ears was the booming voice of the vice-principal. I don't know if I heard what he said, but I knew he was trying to embarrass me, to squash me, and to take my ribbon; and I'm not sure what I said but I know I kept standing and the ribbon stayed on my arm. The igloo of big boys melted away and I marched off on my own, a fire burning in my belly.

But I wasn't alone. Children all over the country were marching and being killed. I was marching with them.

I can't be sure that the Somethings I ended up doing were always the Right Things. I wanted to do Something about the Wrong Things. As I grew from a schoolgirl to a university student, my marching legs got longer, and my intel-lectual fingers grew: I was able to grasp more in the fist of my mind. I nurtured ideas about what was Wrong and Why. Seeds of these ideas were planted by

Reading Stories to Fire Dogs

what I saw (in dusty townships and backyard rooms); and read (in thick books and small pamphlets); and heard (shouted at big meetings or whispered over kitchen tables).

My mind painfully held the sharp realization that many of the Wrong Things were because of 'Apartheid' – which squashed the Blacks to serve the Whites; and that this was part of 'Capitalism'. Under capitalism the most important thing is to squeeze out more and more money for a few people to spend. This means that most of the other people – as well as the earth and sea and all the creatures and plants and *goggas* [2] – have to pay the price.

I jumped into a fast-flowing river of people that pushed against the tanks and banks of Apartheid Capitalism. Even while being swept along, I still wondered what *exactly* I should be doing. I longed for a progressive democratic body that could see clearly into my heart and abilities, and look at the needs and opportunities of the world and then tell me: 'Sally, your job, right now, is ...'

One night I had a dream that I was in a rural African community. Not the prospering village of my romantic fantasies, in which children eat sweet jungle mangoes in front of beautiful mud huts. These people were dust-poor and their red-brick houses were patterned with cracks and crumbles. They had to coax their scratchy vegetable gardens into giving them food. Around the sandy roads and grassy verges strayed some hairy, smiling dogs. Their ribs were showing but their tails wagged easily. The women in this village were organized and had a powerful voice. In consultation with a giant sacred rock, they decided and acted on the needs of the community. These women decided what my job was to be, and came to inform me.

'We need you,' they said, 'to read stories to dogs.'

I was filled with a sense of peace and purpose and set about my task at once.

When I awoke from my dream I was eager to step into my newly discovered role in the revolution. I started with Riska, my respectable, old, Ridgeback dog. She spent a lot of time on her comfortable and mildly smelly couch – and so was a captive audience. Captivated, however, she was not. She listened politely to a few tales of Winnie-the-Pooh and Piglet, but although I was enjoying myself, she displayed signs of awkwardness and even embarrassment. So I let her be, and forgot about my dream for some time.

My dreams were put on hold and my activism was made passive when I was swamped by a heavy illness. Years moved slowly by as I battled with my own body. At the same time the struggles of the Revolution gave way to the handshakes of Co-option and my heart sagged with disappointment.

7

The Fire Dogs of Climate Change

I eventually climbed out of the quicksand of my sickness and watered my heart with hope. When I emerged to solid ground, I awoke again to the struggles and scars on the earth's surface, and the question of what I should do. A kaleidoscope of questions patterned my thoughts:

'Is it only the human race we should be looking after on this planet? What is the biggest threat to the Earth and her creatures? How does the struggle of the workers and the rural poor relate to the struggle for survival of all other life on this planet? Who can voice, and fight for the needs of the Earth?'

I explored these questions with a range of activists and environmentalists, who were open and friendly, informed and inspired, sometimes tired.

Clarity of my role, however, continued to elude me, so I consulted with the Earth herself. I sat at the foot of the mountain, facing the sea, and listened quietly. Under the silence of the clouds and in the rhythmic hushes between the waves – I heard her voice.

'Your job,' she said, 'is to open the heart. To open the collective heart of people.'

This instruction contained no name and address of an appropriate Non-Government Organization (NGO) and did not fit my political perspective of what needs to be done. Which is (obviously) to destroy the profit-seeking, consumerist economic system that is responsible for the destruction, pollution, exploitation and degradation of the Earth and most of the people on it; and that (obviously) my role is to support the poor and the working class in challenging and ultimately wresting power from the capitalist class, and establishing a democratic, international socialist government that would meet the needs of the people and the environment.

I could not really see how this related to opening the collective heart. So I forgot about the Earth's advice... for a while.

Later, in conversation with a new friend [3], the story of my long-ago dream and the Earth's advice surfaced and floated like bubbles on our pool of talk.

He smiled widely and said, 'Don't you know what 'dogs' represent? This is the Year of the Fire Dog!'

He explained that according to Chinese astrology this year is governed by the element of fire and by the animal, the dog. February 2006 to February 2007 is the Year of the Fire Dog.

I scoured some astrology websites and learned that dogs symbolize those who look after the collective interests of the community. They are the watchdogs and guardians of the Earth. They bark to draw attention to injustices. While others sleep, they prowl the night, guarding the grounds.

Reading Stories to Fire Dogs

Astrologers predict that in this year of the fire dog there will be natural disasters because the Earth is out of balance. Dogs will bark fiercely, inspiring debates about how to save the planet, and pushing governments to change their policies.[4]

So! Now that I know what dogs are, I can begin the job I was given – reading stories to dogs. To open their hearts. Stories that help them to feel and understand, so they can think and act, bark and run together in wild places, bare their teeth and, if necessary, bite. Stories of those fighting, and those licking their wounds. Stories for those who are dog-tired, and want to go for a walk in the fresh air and wag their tails.

This is the first story I am reading, and you are the first fire dogs to hear it.[5]

The Fire Dogs of Climate Change

Tails of Integrity

The phone rings. I answer it.

'Nozizwe[1],' he says, 'it's Vincent.'

My heart sinks – it is already so heavy this morning.

'Vincent!' I reply, '*Kunjani*?'[2]

He surely needs my help, but I don't feel able to give attention to this man today.

'Nozizwe, I am so sorry to hear about Riska. That's just terrible. *Ndinolusizi, nyani*.'[3]

My bones soften into grief.

'Today is her last day,' I say.

I allow myself to cry quietly, and him to comfort me gently over the phone.

'Send me a photograph of her,' he says, 'and I will do a painting for you.'

Riska is my dog. She is old and very sick and I am putting her down today. Vincent is an artist and the son of a close friend. He is in Victor Verster Maximum Security Prison – 27 years for armed robbery.

I sit next to Riska, my rust-red Ridgeback, who lies curled in a circle on her couch. Her nose rests peacefully on her paws, but her breathing is heavy. The vet is coming at twelve o'clock.

I feed her little pieces of dried springbok sausage, which she snaps up like a hungry puppy. I sing to her and tell her stories of our times and adventures together. We have had some remarkable and wonderful experiences. A few years ago she won a radio competition for singing dogs that gave us a trip to London. When we came back we took her canoeing down the Orange River.

But it is not the unusual and the special that makes her precious. It is the simple love I feel for her, and she for me. This is what has brought me 14 years of joy and these last few days of sorrow.

Last night the spirits of the wild dogs bounded in the front door to visit her. They had short yellow hair, heads with big ears and sharp-teethed wide mouths. Her young spirit leapt from her swollen body and played with them, her tail dancing wildly. Then the yellow dogs ran away. I left a trail of dried sausage to invite them back, but they didn't come. So I phoned the vet. He is coming at noon.

The Fire Dogs of Climate Change

I help her walk into the front garden and balance her swaying body as she pees. The curves of the mountain embrace us and the wind brings the sound of the sea. The vet arrives. I lie her down with her body on the grassy soil and her head on my lap and hum one of the songs she likes to sing.

He injects her. Overhead, the clouds are silent. He feels for her pulse and checks her breathing, but I already have irrefutable proof that she is gone: I am holding a piece of dried sausage in front of her nose and she is not eating it.

The vet goes and the wild dogs arrive. She goes back and forth between them and me, then follows them down, down and through the hole to the other world, which lies underneath the Life-and-Death tree, which the Bushmen[4] used to make poison arrows. It grows next to the pathway at our front gate.

It is now nearly two years since Riska died, but when our cat, Wuppertal, walks past the base of that tree, she still sometimes rips the air with her claws. Just like she used to do to Riska when they first met.

Now Riska runs with the wild dogs and she also lies curled, snug in my heart. These days my heart is trying to hold the illness and loss of a creature much bigger than Riska.

Earlier this year, the Year of the Fire Dog, I read the first few paragraphs of an article by James Lovelock. He said that it was the most difficult article he had ever written. He was writing as a physician of Gaia, the organism of the Earth. His heart was heavy because of the sad news he had to break to us about her condition. He equated his task to that of a policewoman who has to tell a family that their child has been found dead. My heart sank. I like this planet. I am really attached to life. I was not able to finish the article but it looked like the Earth was not going to slip away easily, with her head on my lap and a sweet song in her ear. She was likely to suffer for ages.

I carried a large stone of grief around in my belly. I did not know what to do with this heavy weight.

A while later I read the whole article. It seems that in spite of the grave introduction, Lovelock's prognosis was not terminal illness – but a hundred-thousand year fever, the extinction of most species and the end of life as we know it.

I was delighted. Is that all?

I thought of my similar response of bizarre delight when, a few years ago, the physician of my nephew, JP, gave us some difficult news.

While surfing, JP had been attacked by a shark. It bit off his right leg and the blood loss was severe. He was declared dead but, after 45 minutes without a pulse, he had been revived. He was in a coma and his brain was swelling. For days our hearts were suspended in the hospital waiting room, fearing imminent death or permanent brain damage. One night we were informed that the doctor was coming to talk to us. When he arrived his face was grim and he requested to meet with only the close family. We followed him into a small room where we all sat down, leaning forward, towards each other.

'There's no easy way to say this,' he began, 'so I'm just going to come right out with it.'

Our hearts twisted and we waited for the heavy hammer to fall.

'They have found a piece of JP's surfboard,' he said, 'washed up on Bikini Beach in Gordon's Bay, with JP's leg attached to the board's leash.'

We looked at each other and laughed aloud, then quickly back at the doctor to check: 'Is that all? Is that what you came to tell us?'

He nodded and looked at us oddly as we collapsed into laughter again.

This hysterical laughter is easy to talk about now because our anxiety and grief ebbed away as JP's strength flowed back into his body.

The future of the Earth still hangs out of balance however. In order to understand the Earth's condition it has been necessary for me to learn more about this planet. In ecology and its related fields, I am a novice. Although we were smattered for some years with biology lessons at school, the only thing I remember with certainty was the biology 'practical' my friend and I invented to verify the existence of peristalsis: I did a headstand, while he fed me chocolate peanuts.

Now I have begun to learn a little about this Earth on whose curves we live. I have lightly touched the surface and I am already struck inarticulate with awe.

A description in the *Gaia Atlas of Planet Management* is better able to find the words to capture something of this wonder:

The sphere of rock on which we live coalesced from the dust of ancient stars. Orbiting round the huge hydrogen furnace of the sun... the globe is white hot and molten beneath the crust: continents ride in a slow dance across its face, ocean floors spread. And between its dynamic surface and the vacuum of space, in a film as thin and vibrant as a spider's web, lies the fragile miracle we call the biosphere...

Within this life realm, every organism is linked, however tenuously, to every other. Microbe, plant, and mammal, soil dweller and ocean swimmer, all are caught up in the cycling of energy and nutrients from the sun, water, air and earth. [5]

The Fire Dogs of Climate Change

Everywhere I dip into the lucky packet of earth-stories I find wondrous tales. I learned, for example, that there is a current in the sea that has a thousand-year circuit, which moderates climate temperature and rainfall around the world; and that algae is not just pond-slime that your dog likes to eat – but a major ingredient in the recipe for creating and sustaining all life.

Lovelock's 'Gaia hypothesis' is that the Earth is a living organism, which, like a body, does its best to maintain its own health. We humans are part of this living, breathing organism. We are made from the same stardust. Like the Earth, we are mostly water, combined with sun and soil and air. In arrangements that are so unlikely and fragile, so creative and chaotic and yet so perfectly arranged.

So impossible – but here we are, here and now. We walk on the Earth without floating off into space, we breathe in and out, and we do it all just *sommer*,[6] as if nothing's really happened.

The organism of the Earth is gravely ill. This diagnosis is backed up by numerous physicians. The cause of the illness is damage done by humans. A few humans more than others. The illness is going to affect all lives on Earth – some more than others.

Life on Earth is an intricate and delicate dance that has evolved over millions of years. Humans are out of step with this dance. Moving to the tune of greed and hunger, we take and burn and kill more than the Earth can restore. We give back rubbish and poisons into the depths of the sea and the heights of outer space.

We have polluted the atmosphere so severely that we have made a bell jar of gases, which is trapping the sun's warmth around the Earth. The rise in temperature and the melting of giant icebergs will throw out of kilter the ancient cycles of currents in the water and air, at a rate that will make it impossible for most species to adapt to. We are moving fast into an era of mass extinction in the life of the Earth. And she is older and more scarred than before.

Does this all sound unlikely or incredible? When you look around you, life still looks okay, right? The people running or managing the Earth would not allow things to get this bad, would they? There are managers, surely? And they can't really be driven only by profit and power? Perhaps you want some proof: facts, figures and references. I gathered a few together and made up a 'Fact Sheet' for you to read.

I typed it up yesterday.

Tails of Integrity

Today I should be gardening. I need to put my feet and hands in the soil. Cut back the overgrown poison-arrow tree that we have to duck under at the gate. I need to feel the wind and the sun, and plant living things and watch them grow.

Last night I slept fitfully between my fact sheets, trying to keep my head above the senseless killings all around me. For the last few weeks, I have been researching the facts and figures on the damage to life on Earth. My mind can see many numbers, but my heart struggles to hold so many feelings, so I push them aside. But now, as my womb lets go its lining, sounds are louder and sharper, and feelings are as clear as bright fish in a lucid pool. They rise to the surface and demand that I feel them. But I just can't.

What can I feel about so many million tons of shark fins – cut off living sharks – so that people can eat shark-fin soup? I could barely hold the loss of one leg severed from my nephew by a single shark.

How can my heart hold millions? The millions who die from hunger or foul water? The millions of dollars made? The millions of species that will disappear? How can my heart carry the unnecessary deaths of so many young and soft-faced children? How can I mourn the loss of the brave beauty of all the wild creatures? And say goodbye to every bright delicate pattern of the growing things?

I am going now to work in the garden. Perhaps the dam of feelings that my heart cannot carry, the river in my womb can wash into the earth. The tides of the moon are forgiving and ever hopeful.

So I prune and plant, my feet and hands in the soil. In the evening, my tall warm man shows me ten tiny chameleons, sleeping. Baby dragons, gripping their blades of grass, glowing in the torchlight, their tails coiled in small sweet spirals. Forty million years old, and just been born.

In the shallow bed of the night, our green-eyed cat finds her place, heavy on my chest and belly. She curls her thick black tail around the curve of her body. I lie deep into darkness while her purr thrums through my heart down to my toes.

I know it is important to understand the nature of the Earth's illness so that we can support her healing. This story, however, is not about diagnosis nor cure, but the wrestle with my heart. How can I face the dire facts of the condition of the Earth without leaping in despair off the melting ice-cliffs, or burying my head in the hot sand of denial?

How can I respond with integrity?

When I opened my heart softly, and listened quietly, I found something.

Here it is: whether the Earth is thriving, mildly ill or terminally diseased

The Fire Dogs of Climate Change

should not make much difference to how I relate to her. I should know I am part of her and act to keep her healthy. Give her love and care, not poison and abuse. Life (like love) is forgiving and hopeful and, like a bold flower pushing out from the narrow crevice in a rock, will take any gap we allow. However small or large.

When my dog, Riska was young and fit, I did not kick her just because she was strong enough to handle it. When she grew older and slower, I did not find reasons to neglect her. Instead I declared she was in her golden years and took her on wild adventures. She came with us down the rapids of the Orange River wearing a rainbow-coloured life jacket, and to Namibia where she quivered and howled with primordial excitement to see a charging *gemsbok*.[7]

When all my attempts to keep her alive had failed and I knew she was dying, I did not ignore or abuse her, but sang her songs and fed her dried sausage. Right up until she left to join the spirits of the wild dogs. And so it should be with the Earth. However strong or weak she is. However much hope there is or isn't. I can love her, heal her, celebrate and appreciate her. I need to do this not from despair, fear nor shame; but because by so doing I experience my own integrity.

Integrity is being true to myself and others, whether I live in a web of damage, or a climate of joy. May I be as true and expressive as the dancing tail of Riska.

Integrity is being integrated and knowing of what I am part. Like the black curve of my cat's tail is part of her, so am I part of the vibrant layer of life that hugs the curves of the soil. I am integral to the Earth.

The Earth, like the spirals of a chameleon's tail, moves in cycles, through boundless space and endless time.

Vincent, from his prison in Paarl, did send me a painting of my dog Riska. She is lying curled in a rock hollow, made long ago by water, on a huge rust-red desert rock, the colours and curves similar to her own, the blue sky above her. His painting shows the tip of her tail inside her own mouth. Like the most ancient symbol of the snake swallowing its tail – the infinite and eternal cycles of the universe. I hold this circle of her snug in my heart – the shape of integrity.

Fact Sheet 1

The Problem – Environmental Destruction and Climate Change

The Earth is facing an environmental crisis

Over the last 100 years, the Earth's ecosystems and resources have been damaged and used up at a rate faster than they can recover. Seventy percent of all the Earth's hotspots have been destroyed. 'Hotspots' are areas that are the most biologically diverse (e.g. coral reefs and rain forests), and they are crucial in generating and sustaining all other life. Half of the world's coral reefs have been destroyed (bleached) by rising sea temperatures.

We are entering the 'Sixth Great Extinction' in the life of this planet. We are losing approximately 30, 000 species a year, which is 120, 000 times greater than the historical natural rate of extinction (see Anderson, 1999; Millennium Ecosystem Assessment Report, 2005). As global warming increases, this figure will rise dramatically.

The destructive practices of humans

The environmental crisis is caused by certain human practices. These include: the destruction of wilderness habitat (through human constructions, developments and agriculture); introduction of invasive alien species; overexploitation of natural resources; and the creation and use of toxins and pollutants. One of these pollutants is carbon dioxide (CO_2) (Millennium Ecosystem Assessment Report, 2005).[1]

The main cause of climate change is CO_2 released when coal and oil are burned to make energy

The Earth is getting warmer. This is because, over the last 200 years, we have released pollutants into the air called 'greenhouse gases'. These gases trap too

Fact Sheet 1: The Problem

much heat in the atmosphere of the Earth, causing a progressive rise in temperature. This is known as the 'greenhouse effect' and it causes global warming and climate change. Carbon dioxide (CO_2) is currently the main greenhouse gas (GHG); it made up about 75% of the Earth's GHGs in 2000. It is released when fossil fuels – coal and oil/petrol – are burned to provide electricity and transport. It is also released when indigenous forests are destroyed. Globally, deforestation of tropical forests is occurring at a rapid rate (an area equivalent to 20 rugby fields per minute). This causes about 18% of the Earth's GHGs, and is extremely destructive to the environment.

The other GHGs should not be discounted because they are in the lower percentages. Most of them are even more destructive than CO_2. Methane, for example is a very potent greenhouse gas that already constitutes about 14% of all GHGs and will be released in large volumes from the land and sea once global warming reaches a 'tipping point'. This may cause 'runaway' effects, resulting in rapidly escalating and highly destructive global warming and climate change.

(www.climatenetwork.org)

Climate change causes natural disasters, deaths and extinctions

Global warming affects major weather-regulating currents in the sea and air, and causes climate change. Effects include: melting ice caps; rising sea levels (submerging coastlines and islands); increased 'natural' disasters (e.g. floods, tidal waves and hurricanes); changing rain patterns; extended droughts; crop-failure and disease. The Intergovernmental Panel on Climate Change predicts escalating famine, wars and numbers of refugees (www.ipcc.ch). Southern Africa will be hit particularly hard. Although it is the rich who are responsible for most of the GHG emissions, it is the poor who will suffer the most.

We are already experiencing these problems, but they will get progressively worse, increasing in correlation to the level of previous GHGs emitted.

We are damaging the lives of our children and grandchildren

Climate change has a delayed effect. The climate changes we are experiencing today were caused by emissions in the 1970s. Today's CO_2 emissions will have an effect in about 30 to 40 years. Often humans are only motivated to act once things get really bad. We find it hard to look ahead. In this case we will be leaving future generations to face the consequences of our actions (and inactions).

The Fire Dogs of Climate Change

We need to keep global warming below two degrees Celsius

If global warming increases more than two degrees Celsius (2° C) above pre-industrial levels, the cataclysmic events mentioned above will accelerate. Thirty percent of all species may become extinct. A 2° increase could also precipitate runaway warming effects (as methane is released from the ocean). If global warming reaches 6° we are likely to lose 95% of all species, and threaten the basis for most life on Earth.

If we do not make major changes now, we will definitely reach 2°, and possibly 6° by the end of this century. The actions we take within the next ten years will determine the future of life on Earth.

A minority of people and corporations cause the most environmental damage

The rich countries, containing 15% of the world population, account for about half of the GHG emissions. Twenty percent of the people on this planet, mainly from Western economies, is responsible for 90% of its consumption of resources. While this minority gets too much, the majority does not get enough: about 40 million people die from hunger each year. Many of these deaths are in countries with food surpluses. This reflects problems with our political, economic and value systems. Most of the environmental damage is the result of economic and social practices that are designed to benefit the wealthy minority of the Earth's population.

Figure 1 (*page 57*) shows a map of the world reflecting the amount of CO_2 emitted in each country. The bigger the size of the country on this map, the higher their emissions.

Figure 2 (*page 58-59*) gives a breakdown of the different sectors (not classes) that produce GHG emissions worldwide. It also charts the activity or 'end use' related to the emissions, and the type of GHG emitted. This format is useful in noting the difference between energy, CO_2 and GHG emissions. This 'spaghetti chart', if studied carefully, is worth a dozen 'pie charts'.

Figure 3 (*page 60*) is a reflection of the per capita CO_2 emissions (i.e.: 'per person' emissions).

The poorest one billion people in the world are responsible for three percent of the CO_2 produced. Some of the countries with the biggest emissions (e.g. US) have been the most resistant to reducing their emissions. Big businesses (in particular the oil compa-

Fact Sheet 1: The Problem

nies) are making a lot of money from the industries that are producing the GHGs. All over the world, businesses pressurize national governments to protect their profits.

Who and what produce the most greenhouse gases?

It is important to establish who and what is responsible for GHG emissions in order to tackle the problem effectively. The energy-supply industries use coal and oil to make electricity and fuel, and are the biggest overall contributors to GHG pollution (61%). All over the world, the multi-national fossil-fuel industries, and the governments that support them, are responsible for much environmental and social damage in addition to CO_2 emissions (e.g. US wars on Iraq and the destruction of the Niger Delta).[2]

For decades, these industries and governments have suppressed and resisted the replacement of oil and coal with healthy alternatives. Internationally, big corporations (industrial and agricultural) use most of the energy and emit the bulk of greenhouse gases. The wealthy classes also use a lot of energy in their homes, methods of transport (cars and planes), and in the products they consume. The poor have a very low carbon footprint.

As you can see in figure 2 (*page 58-59*), electricity and fuel are the biggest problem; followed by deforestation and agricultural practices. The chart reflects GHGs at a global level, and there will be some differences between countries. In the developed countries, for example, there is a greater percentage of GHG emissions from the transport sector (e.g. US is 27%). In the developing countries there is a greater percentage from agriculture and deforestation. Note, however, that many of the emissions in the developing world are related to 'feeding' the developed world (e.g. chopping down forests to grow ingredients for fast food or bio-fuels).

Although this is the most comprehensive single chart I have come across, it does not provide every aspect of the statistics. For example the GHGs from the transport industry cannot be reduced to just those listed under 'transport', as this refers to 'tail-pipe' emissions only. There are also emissions involved in the production, marketing, servicing and destruction of vehicles, and the building of roads, runways, parking lots etc. (which may be charted under 'industry', 'commercial buildings', 'iron and steel', 'cement', etc.). Nor do these figures show how much of the transport is for private (and industry) use, and how little is public transport.

Similarly, although 'food and tobacco' appear low on the energy 'end use' chart, this does not take into account the significant emissions from other areas that are related to the production, sales and consumption of these products (charted as 'agriculture', 'transport', 'commercial,' etc.).

The Fire Dogs of Climate Change

It is crucial to have clarity on the nature and causes of our climate problems. We need to face up to them: see them, know them and feel them. This may leave us feeling overwhelmed or depressed, but if we ride through this, we can find solutions. We can find hope in our hearts, and we can take practical actions that really do make a difference.

Riding the Moon

It is early Saturday morning, still dark and I am half-asleep. The traffic has stopped and the wind is still. My cat is sleeping too deeply to purr, and my man is curled quietly on his side, his warm body soft in sleep. It is now that I can hear the sea. Its gentle roar fills the room. As I lie on my back, I allow myself to relax deep into its sound. The firm mattress melts away and I am floating on the rhythmic waves of the ocean. My body disappears and I am just a gentle movement, pulled back and forth, riding up and down. I sink into a deep sleep, and float between the seaweed of my dreams. I awake to the sound of hooting and find myself high and dry on my bed. The cat stretches and Bowen yawns and pulls me against his chest.

'I dreamt I was the sea,' I say. 'I was riding the moon.'

We have all come from the sea, and our bodies are mostly water. Since our oldest ancestors first crawled out of the sea and onto the land, people have felt the pull of the ocean. We flow to the sea to feed our bodies, minds and hearts. Just as the ocean currents regulate the temperatures and weather around the planet, so can the sea help us to mediate the ebb and flow of our hearts.

The moon moves around the Earth, pulling the water that is in the ocean and in my body. When I ride the rise and fall of the tides, I am riding the moon.

I try to explain this to Bowen.

'Do you know what I mean?' I ask.

'Mmmmf,' Bowen replies.

'The moon was a dolphin leaping,' I say.

He smiles.

'My dreams were seaweed,' I say, pressing my face into his neck.

'Wee Swagger,' laughs Bowen, 'and the kelp.'

I know this story of Bowen's and I smile. He had met Mr Wee Swagger when he went to Camps Bay beach with his friend Craig.

Bowen and Craig strolled past oiled sunbathers lying alongside bright umbrellas and walked beside the low waves crashing on the shore. In the shallow clear

The Fire Dogs of Climate Change

water near the rocks, they saw a small black shark gliding across the rippled white sand.

They wanted to swim but had no costumes or towels. Craig had underpants but Bowen did not. So Craig lent him his white T-shirt, which Bowen turned upside-down. He put his legs through the armholes and tied the shirt around his waist like a nappy. These two tall, bearded men ran into the sea, splashing and swimming and holding onto their not-well-elasticized swimwear.

When they got out they were wet and cold, so to warm up they played frisbee with muscle shells; flinging and running and diving in the sand to catch the spinning black and pearly-blue shapes. While doing this they noticed a man further down the beach who had swimwear of a similar brand to theirs. This man, who was large, brown and pot-bellied, was wearing Y-front white underpants with tiny, vest-like holes in them. They watched him swagger majestically along the shore in front of all the sun-tanners. He would pause his swagger every now and then to face the sea and make a long arc of a wee into the waves.

Craig and Bowen went over to make the acquaintance of Mr Wee Swagger, who seemed merrily drunk and was happy to pass the time of day with them. He noticed they had been swimming and praised Bowen's fine swimming apparel.

'You whites came from the sea,' he said, 'we came from the land. You must teach me how to swim.'

They came to an arrangement for swimming lessons. Mr Wee Swagger was to go into the sea, holding onto one end of a long, strong piece of kelp and Craig and Bowen were to stand on the solid shore and hold the other end, while issuing instructions.

Mr Wee Swagger, tightly gripping the thick brown kelp, backed out gingerly into the mighty ocean, interjecting his passage with shouts of protest and delight. Bowen and Craig, though disabled by mirth, managed to hold onto the other end of the kelp. Mr Swagger bravely continued as his knees, then his round belly, were submerged in the waves. When he was up to his armpits in water he awaited further instructions.

At this point, out of nowhere, bounded a pitch-black Labrador. It leapt long and accurately through the air and ripped the kelp from Mr Swagger's extended arms. Clearly petrified, he could hardly shout for help. The two men plunged in and saved him.

The dog raced up and down the sand dragging the kelp, entreating the men to continue with the game. But Bowen and Craig could not convince Mr Wee Swagger to play some more.

I laugh as I recall Bowen's story, but I know that it is not only for fun that we go to the ocean. I once had a grief that my body could not bear. I took it to the sea.

Every year on the anniversary of my loss, the pain of my grief would return. It would start in my belly, rising like a black oil-slicked bird, its beak open, screeching silence. Sobs would shudder through me, stealing all my breath and emptying me of tears, which, if I were lying down, would fill my ears.

My heart and mind found some resolution but my body was unable to bear the grief. I needed help with carrying its weight. I collected my tears in a small white cloth and headed for the sea. I travelled along a snaking sandy path, my footsteps chasing small lizards into the low green scrub. I walked down towards the wide dark-blue water that lay under the sky.

I stood on a rock at the top of a jagged cliff. The waves were churning softly around the rocks below. I spoke to the sea and asked for its help. I dropped my small cloth down through the sea-spray and into the swirling white-blue patterns below. It was swallowed by the water.

I felt the full breadth and depth of the ocean around the sphere of the Earth, back billions of years to the beginning of life, across all the passing lives and deaths, the endless waves of swimming joy and quiet losses of exquisite creatures with fins and fronds, tentacles and wings, colourful and transparent, tiny and huge, coming and going. There is nothing the ocean has not seen.

Onto the broad and deep shoulders of the sea fell my little white cloth. And the sea was able to bear it. It could do more than bear it. From above the sound of the swirling waves I heard a short hard splash, then another. I looked up to see a whale breaching. A huge dark form rising up, leaping and crashing; again and again; pushing a song of joy up from my belly, pealing through my heart.

On my return from the sea cliffs I passed a zebra standing still on the *fynbos* [1] hillside. Its stark black and white stripes accentuated the crescent curves of its round, pregnant belly.

I ache from the knowledge that the deepest losses that the ocean has to hold are those caused by humans. The sea has helped me to bear this pain too. On another pilgrimage to the ocean, I joined an expedition to swim with dolphins in Mozambique. There are many stories about dolphins rescuing and healing humans. Their sonar call is so sensitive that they can respond to the heartbeat of a human foetus inside its mother's womb.

On the evening of our arrival we prepared for the boat ride we were to make in the morning. We were shown how to put on a snorkel and a life jacket correctly, and what we could do to encourage the dolphins to stay and play. We also learned that the toxins that humans have washed into the sea, concentrate in the dolphin mother's milk. Her firstborn suckles this toxic milk until its heart stops beating.

That night I stood on the beach, the dark starless sky smudged with clouds, and listened to the crashing of the ocean. The sea we come to for healing. The dolphins we dare to ask to play with us.

Into the channel made by the small body of a newborn dolphin, flowed waves of knowledge of the damage that humans are doing to the ocean. The hundreds of dolphin corpses washed up on the beaches of Zanzibar, [2] the huge dead zones in the sea caused by agricultural and sewerage pollution, the over-fishing, the nets and the killings.

A hot cloud of shame moved over me. I didn't own the industries that do the worst damage but I probably used their products and I was not fighting their practices. I stood small and alone on the open dark beach. I did not know how to be responsible to this vast and generous sea. A heavy weight settled on my shoulders and I crumpled down onto the sand.

In the morning, wearing an orange life jacket and a heavy heart, I boarded the speedboat that pushed an arrogant path through the big waves and into the open sea. In the low swells of the clear deep ocean appeared the swift smooth shapes of the dolphins. I looked down, and next to me, shadowing the boat, was the blue-grey form of a speeding dolphin. My heart shook to see such grace.

They swam away from us and into a long low hill of water that was a wave crashing onto the shore. They disappeared. Then out of the translucent green curve of the back of the wave we saw their dark crescent arches shooting up into the air; leaping against the force of the great breaking waves, high into the sky.

Splashes of laughter surged up from inside me, pushing through the heavy weight on my heart and shoulders.

Bowen gets up out of bed and our cat, Wuppertal follows him, padding along the wooden floor to the kitchen to wait impatiently for her breakfast. I sit up in bed, but do not want to move, still wrapped in the seaweed of my dreams. I want to sit until I have surfaced. I want to grasp this message from the sea, turn it over like a shell in my hand.

The sea is showing me how to feel and to flow. Sometimes I have wrapped a numb wall around my heart and closed my mind to keep myself 'safe' from the damage around or inside me. But when my emotions and thoughts are closed, I block the flow of all knowledge and feeling.

Riding The Moon

The sea rides the tides pulled by the moon. If I open to the rise and fall of my sea of feelings, then the full waters of life flow through me. I can move with the joy as it leaps up through the waves. Happiness can flow through my veins and wash away the heaviness in my heart. When I ride the ebb and flow of the wide deep sea then I am able to think and feel. With this comes the ability to respond. Response-ability.

I remember the dream that I had just before the car hooted me awake. I was floating peacefully on a flat calm sea. Close by I could hear soft explosions of breath as huge gentle beasts shot fountains of water into the air. I heard playful splashing and saw all around me wet, dark, curved backs, arching out and into the water. Dolphins, I thought. Then one or two of them raised their heads – and I saw the long ears and black snouts of the dogs. Sea dogs! Swimming and diving through the water.

To be response-able to the sea, we need to be sea dogs. Guardians and watchdogs that bark and fight the destruction. But who also swim and leap with joy.

I get out of bed and prepare for our visit to the sea dogs who are docked in the Cape Town harbour.

Bowen and I are welcomed up the wooden gangplank onto the Sea Shepherd boat, the skull and crossbones flying on a black flag. The crew are people riding the waves of their responsibility in a small and sturdy ship across the wide oceans, destroying illegal fishing nets and defending endangered whales, dolphins and seals from attack.

They take us into the boat's black metal curves and across its long wooden deck and we explore small cabins and big maps, ropes and ladders, lifeboats and water cannons. The smell of rusty salt and engine-oil is replaced with spicy tomatoes as we enter the warm galley of the ship. Over supper we hear stories of deep stormy seas, battles and sinking ships, the deaths of giant gentle creatures and the fights with their killers.

We visit the wheelhouse, where under a glass bell jar is a gift from the Dalai Lama to the captain of the ship: a wooden gargoyle, its wings big, and its eyes and beak sharp – a symbol of compassionate wrath.

On the foredeck we are shown a big copper bell, which is rung when dolphins are sighted. On hearing its clanging peals, the whole crew will run to the front deck and laugh and smile at the dolphins riding the waves made by the surging bow of the boat.

The Fire Dogs of Climate Change

We leave the sea dogs and travel out of town to visit my brother for dinner. I sit on his wooden deck on the bank of a tidal river, the sound of the sea in the distance. Swallows and bats swoop down to the surface of the flowing water. Two men play guitar and mouth organ and sing beautiful Afrikaans love *liedjies*.[3] They sing about the taste of the sea on your body. My two-year-old niece comes to me in her pyjamas and states: 'I want to dance!'

I lift her into my arms and onto my hip and together we sway and twist and turn to the music. The flatness of the evening sky gives way to a deep blackness. The layers of dark night are splashed with fiercely burning stars. We dance to the song of the guitars and the bright stars and the tides of life flowing through us. We sing and twirl and her laughter peals like a bell and she shouts the names of the friends she loves most.

In the sky we see the crescent moon rising. The glowing curve of its back is arched as it leaps up from the deep black sea of the night.

'I want to ride the moon!' she calls out, 'I want to ride on its back!'

'So do I,' I say.

Fact Sheet 2

The Solutions – Energy Efficiency and Renewable Energy

Mitigation and adaptation

We need to mitigate (i.e. reduce the effects of) climate change. However, no matter how swiftly we act, we are still going to experience some serious climate change (because of previous emissions). Hence the need for climate change 'adaptation'. This book, however, focuses on mitigation: I find it hard to discuss the life boats and inflatable jackets (however crucial they may be) when the ship we are on is still heading straight for that iceberg. Let's turn the steering wheel, so we can change course and avoid the worst collision. I find it difficult to give my full attention to disaster management, when we have not yet done our best to minimize the disaster.

But there need not be such a sharp distinction made between adaptation and mitigation. There is a lot of overlap between them. Environmentally sane practice could simultaneously address the needs of both mitigation and adaptation. For example, decreased consumption by the rich, building ecosystems and social systems that support the poor, a low-carbon economy and renewable energy are all crucial to mitigation strategies, but they also build 'climate resilient' communities (i.e. are part of adaptation).

Both mitigation and adaptation need to be addressed in the context of supporting and building a healthy natural environment. Short term band-aids that create further environmental damage (e.g. nuclear power) are only going to boomerang troubles back onto us and our children.

We must find mitigation and adaptation methods that reinforce each other and support the natural environment. For example, restoring natural ecosystems (e.g. wetlands, mangrove swamps and forests) reduces CO_2 emissions (mitigation) *and* assists with climate change adaptation (floods and tidal waves are reduced by wetlands and mangroves, and harsh weather moderated by trees). These ecosystems are the basis for all our food, water and other survival resources.[1]

Fact Sheet 2: The Solutions

Targets must ensure that the average global temperature rise is below 2° Celsius

Our current concentration of carbon dioxide is about 384 parts per million carbon dioxide (ppm CO_2). Our concentration of *all* the greenhouse gases is measured in ppm CO_2 *equivalent* (CO_2**e** or CO_2**eq**). Greenhouse gas concentration is currently at about 440ppm CO2e. It is rising at approximately 2ppm CO_2e per year. Do not skim over these numbers and terms: they make the world of difference. Although the science is clear, the estimates and numbers can be presented in a confusing fashion. There is a big difference between aiming for 400 and 450 ppm CO_2e; between CO_2 and CO_2eq; and current levels vs. 1990 levels as a reference point for reductions.

The Stern Review (2006) gives the impression that keeping below a 2°C increase (of pre-industrial levels) requires that we stabilize at 450ppm CO_2 equivalent. This means global emissions would need to peak within the next ten years and then fall at more than 5% per year, reaching 70% below 2006 levels by 2050.

In line with this, many climate change professionals and activists (e.g. United Nations and Greenpeace) argue that we must not exceed 450ppm CO_2e. 450ppm CO_2e translates to roughly 350 ppm CO_2.[2] This is the figure called for by the organization, *350*, as well as *Friends of the Earth* (the biggest international network of environmental organizations).

However, 450ppm CO_2e only gives us an estimated **50/50 chance** of avoiding the 2° rise. To achieve a 90% chance of stabilizing below 2° increase, we need to *stabilize* at 400ppm CO_2e or lower (Hare and Meinshausen, 2004; George Monbiot, 2006). We are already higher than this target, but we can bring our GHG concentration back down. Anything higher than 400ppm CO_2e (even slightly higher) significantly increases the risk of extremely dangerous global warming 'feedback loops'. There is already a 20-30% chance that we will face a 2° increase. We need to err on the side of caution.

The United Nations' Human Development Report (HDR 2007/2008) states that if the world were a single country, we'd need to cut down to 50% of 1990 levels by 2050. This would require rich nations to cut (1990) emissions by 80-95% by 2050, with cuts of 25-40% by 2020.[3] HDR argues that developing countries should peak before 2020, with cuts of 20% by 2050.

I would argue that developing countries can and should have higher and sooner targets. [Though tardiness in their commitments should not serve as an excuse for developed countries (e.g. the US) to justify their own inaction.] Commitments from developed countries will not only result in lower emissions, but will also ensure that these countries are not left behind in the renewable energy (RE) market. If lowering emissions requires costs they cannot afford, then the developed countries should assist with these costs.

The Fire Dogs of Climate Change

All of these targets are gambles, calculated to ensure the odds are slightly in our favour. We are gambling with our children's future and the future of all life on Earth. We should see these targets as the absolute *minimum* we aim for. We should increase them as much as we can and as soon as we can – within the parameters of environmentally sustainable practice.

It is essential that we stop our increase in emissions (i.e. 'peak') as soon as possible. 2012 would be relatively safe, 2020 is risking exceeding the outer limits.

Al Gore's bold call for 100% of electricity to come from clean, RE sources within the next ten years, is the target we should be setting for the whole world.[4] However, we should extend this further to cover all GHG emissions, including those from transport, deforestation and agricultural land use.

'Development' and developing countries

The Kyoto Protocol exempts developing countries because of their historically low emissions – allowing them to increase their GHG emissions (at least until 2012). However, the emissions of some of the developing countries will soon eclipse those of the developed.

With the growth of globalization, many industries and multinational corporations are emitting GHGs across the globe. To divide the emissions into individual country emissions can be artificial and misleading. Many of the GHGs emitted in developing countries are to provide products for export to developed countries. To address the global environmental crisis, *all* countries need to act now to reduce emissions.

The historical and ongoing exploitation of the poor by the rich, and the developing countries by the industrialized countries, should not serve as an excuse for the developing countries to pollute and exploit the Earth and its resources.

The vested interests of big business (and the governments they ally with) love to use 'the poor' as an excuse to continue polluting, but in truth it is their own pockets that benefit from the unhindered burning of cheap fossil fuels.

Renewable energy creates more (and safer) jobs than the fossil fuels industry, so should be advocated even more strongly in the developing countries, to alleviate unemployment and poverty.

We do need international 'concessions' to developing countries, but these should be in the form of financial and technical assistance (for renewable energy and energy efficiency practices) – rather than a license to pollute.

Countries that do not engage fully with renewable technology now, will be left behind and their economies will ultimately suffer. There are many marvellous climate change mitigation inventions and practices coming from the developed countries (see for e.g.

Fact Sheet 2: The Solutions

'Inspiring Examples: Groundbreaking Technology', page 78). The developed countries could balance some of their 'ecological debt' to the developing world by supporting the production of these technologies, and accelerating job creation in the poorer countries.

Our Western Capitalist approach to development has failed. We cannot continue with 'growth' and 'development' that is driven by profit, rather than the needs of this planet and the life (including people) that lives on it. Future development should be implemented in a sustainable fashion rather than by repeating the errors of the developed countries. If the developing world were to 'develop' on par with the industrialized countries, we would need nine planets to provide the necessary resources (UNDP, 2007). We only have one.

Governments need to regulate and act fast

From local to international levels, governments urgently need to initiate and enforce targets, laws, regulations, incentives, penalties and education to mitigate climate change. The most viable method to do this is via reduced energy usage, energy efficiency and renewable energy. We also need to address GHG-emitting agricultural practices; and stop the destruction, and support the regeneration, of indigenous forests. These actions need to be regulated and enforced internationally.

We need to develop responses that are sustainable and that support the Earth's environment. Many of the proposed solutions to climate change are damaging to the Earth, and create further environmental crises.

Nuclear power: an unviable response to climate change

Promoters of nuclear power have used climate change to try to resurrect this technology. In the panic of the climate crisis, there are even cases of environmentalists arguing we may have to resort to nuclear power as an emergency measure.

However, we should not let a crisis blind us to the truth about nuclear power. Although there have been developments in nuclear technology and spin-doctoring, it remains an expensive and dangerous option. There is damage and risk to people and environment involved in the mining, processing and transport of uranium. There is the unsolved problem of nuclear waste – which remains toxic for tens of thousands of years. There is the risk of nuclear weapons proliferation as well as the potential hazard of serious accidents.

The process of producing nuclear energy is itself energy-intensive and inefficient. Nuclear fuel is also finite (unlike renewable resources) – so is not a sustainable energy source.

The Fire Dogs of Climate Change

The Austrian government commissioned a detailed and rigorous scientific study into the advantages and disadvantages of nuclear power in the context of the climate crisis. It concluded that nuclear power was not the solution. It stated that, 'even if one were to overlook all (the) drawbacks, a nuclear power scale-up would come too late to contribute significantly towards the solution of the challenges of climate change.'

The report also showed that 'renewable energy sources are superior both ecologically and economically' (*www.lebensministerium.at*).[5] It remains something of a mystery to me why people (apart from those with vested interests such as nuclear scientists, investors and the military) continue to advocate nuclear power. To me the choice between coal and nuclear is like asking, 'Would you rather be poked in the eye with a hot stick or hit on the head with a brick?'

My answer is: 'Neither, thanks!'

Contentious solutions to climate change

There are a number of contentious solutions presented as if they are 'the answer' to the climate crisis. Many of those being pushed by big business, and the governments that support them, are environmentally and socially destructive.

Biofuels

Biofuels (or agrofuels) are renewable fuels made from organic matter (biomass). They include ethanol and biodiesel. They burn 'cleaner' than fossil fuels, producing less 'tail pipe' emissions. Big businesses and governments (especially those of the US and UK) have touted biofuels as a major solution to emission reductions, and many governments have set percentage-use targets for biofuels. According to Oxfam International (2008), rich countries spent US$17 billion supporting biofuels in 2007.

But do they really result in a reduction of emissions? The short answer is 'No' – not if you take into account the CO_2 emissions involved in growing and processing the fuel. A study by scientists found that ethanol made from maize (corn) requires 29% more energy input than the energy generated, whilst ethanol from grass or wood mass requires 45% and 57% more energy input respectively (Lang, 2005).

The large-scale bio-fuel crop production is fraught with problems and presents many environmental and social impacts. Even if we gave over most of our arable land to biofuels, they would only produce a tiny percentage of our energy needs. And we don't want to convert our food-farming land: biofuels compete with food production and have caused worldwide food shortages and price increases. A 2008 report shows

34

Fact Sheet 2: The Solutions

that biofuels have caused a 75% increase in global food prices. The US and EU drive for biofuels has had the greatest impact on food prices and supply, and the current food crisis has pushed 100 million people below the poverty line, resulting in food struggles and riots around the world (The Guardian, 4 July 2008).[6]

According to the World Rainforest Movement (WRM), the amount of cereal needed to fill a tank of almost 100 litres, once, is sufficient to feed one person for a whole year (WRM, 2007).

In many areas (e.g. Indonesia), clearing forests to grow bio-fuels further adds to CO_2 emissions. In other areas (e.g. Brazil) people are being forced off their land to make way for biofuel agribusiness (mainly sugar cane plantations for ethanol).

Monbiot (2008) argues that apart from used chip fat, there is no such thing as sustainable biofuel. It may be possible, however, for small-scale, localized bio-fuels production to be sustainable. I would also look into algae farming (for biodiesel), and the use of waste products (particularly from sewerage and agricultural waste to produce methane and ethanol).[7]

Natural Gas

Natural Gas (i.e. propane, butane, low pressure gas) has lower CO_2 emissions than coal and oil. However, mining and transporting this gas is energy-intensive, and gas supplies are limited. Nevertheless, it could be a useful interim solution, as we cross over from fossil to renewable energy.

Carbon capture and storage

Carbon capture and storage is a process in which CO_2 fumes are captured, turned to liquid under pressure and stored underground. This dramatically reduces the CO_2 released into the atmosphere when coal is burned. However, many argue it is not a long-term solution to the energy problem. It is an expensive technology that is still in its infancy, and will not be available for at least ten years. In spite of this, it is used by the industry to justify the building of new coal power stations. It does not make sense to invest such large amounts of money in developing this technology, when we should be moving away from our coal dependency.

There are also potential risks associated with this process, which has uncertain environmental impacts (some, e.g. Flannery, argue that they are detrimental). Carbon capture does not address the other environmental and social damage caused by coal mining and use. Greenpeace argues against it, stating that it undermines other effective climate change mitigation.[8]

The Fire Dogs of Climate Change

Viable solutions: energy efficiency, renewable energy and a low-carbon economy

In order to reduce CO_2 emissions we need to reduce and avoid the use of fossil fuels. We need to produce and consume less, and practice energy efficiency. Our energy supply needs to come from the one area that promises the most potential benefits with the least potential harm - renewable resources (e.g. sun, wind, water, geo-thermal). We also need to alter the structure of our economy and society (and probably our hearts and perspectives) to achieve a low-carbon economy. Emission reductions need to be enforced in internationally binding agreements. One way of doing this is through the 'cap and trade' system.

International binding agreements: cap and trade?

It is vital to have internationally binding agreements that enforce the reduction of GHG emissions in every country in the world. The next crucial international agreements will take place in Copenhagen in 2009. We need to ensure the negotiators set targets and apply mechanisms that meet the needs of the Earth, rather than the desires of the fossil fuel (oil, coal, nuclear) industries. No easy task.

The Kyoto Protocol was set up by the United Nations Framework Convention on Climate Change in 1997. Most developed countries are signatories [the US has, to date, refused, but Australia has at last signed the protocol (2008)]. It commits these countries to an average 5.2% reduction of 1990 emissions by 2012. One of the mechanisms to achieve this is 'carbon trading'. Emissions trading is also known as cap-and-trade. Under the Kyoto Protocol, developed countries agreed to emissions allowances, i.e. 'caps' on their emissions. Governments enforce this in the private and public sector. Both public and private entities may trade in 'carbon credits.' Companies can then sell or bank these credits, or implement 'carbon-saving' projects in the developing countries, which will give them new 'carbon credits'. This is implemented through the Clean Development Mechanism (CDM).

There are some problems with the Kyoto Protocol, and the practice of carbon trading. Firstly, the reduction targets do not go far or deep enough to avoid the dangerous 2° increase. Secondly, it does not look like even the limited reduction targets will be met. In most countries emissions continue to increase. The carbon trading has allowed big companies to avoid the unavoidable, i.e. the fact that they have to reduce emissions *at source*.

Another problem is that carbon credits in developing countries are acquired in a manner that is quickest and cheapest to the company. This may not support what is

Fact Sheet 2: The Solutions

actually required in that country for large-scale emissions reduction. This is particularly true in off-set mechanisms like the CDM. This project-based mechanism reduces emissions relative to business-as-usual in a developing country, but each ton 'reduced' (or rather, off-set) allows a ton to be emitted in a developed country (that has paid for the carbon credit). At best, the CDM leads to zero increase in emissions rather than a decrease of emissions; if its rules are not carefully observed, an increase will occur.

Some 'carbon-reduction' projects are damaging to social and environmental systems. Hydro-electric dams and agri-business forests are the most notorious examples.

In addition to the official carbon-trading industry, there has been the flowering of companies providing 'voluntary carbon-offsets' to consumers who have a green conscience. Although some of these companies may support good initiatives, others have misleading promises and dodgy projects. 'Offsets' allow companies (e.g. airlines) to put the responsibility onto the consumers, so they (the company) can continue with 'business as usual' (i.e. increasing emissions), rather than cutting down on the emissions that they are producing.

Some activists argue that carbon trading is not only flawed, but also detrimental.[9] A 'Corner House' article states that:

'Carbon trading proponents often assert that trading is merely a way of finding the most cost-effective means of reaching an emissions goal. In fact, carbon trading undermines a number of existing and proposed positive measures for tackling climate change. These include the survival and spread of existing low-carbon technologies, movements against expanded fossil fuel use, and well-tested green policy measures. Carbon trading also undermines public awareness and political participation, as well as creating ignorance.'[10]

There are numerous ways of providing 'sticks and carrots' to penalize fossil fuels and incentivize RE and EE. Carbon trading and carbon taxes are potential ways of leveling the playing field between the 'cheaper' fossil fuels and the renewable sources of energy.

It is important to have legally binding systems in place that will hold the worst polluters accountable for providing the financial resources required to meet the necessary targets. Market forces alone cannot be trusted to provide the necessary reductions. Businesses are in business to make profit, and are notorious for cutting corners to do so. Although we rely on governments to make the necessary laws and actions, they are strongly influenced by large corporations. We need independent bodies (including representatives of climate change scientists, environmentalists and community organizations) with the power to ensure the implementation of the emission reductions.

The Fire Dogs of Climate Change

We need to channel existing CDM and carbon-trading investments into effective and environmentally sound projects; but more importantly we need binding international agreements to ensure that there is a significant reduction of CO_2 emissions *at source*.

So, although carbon trading may have a role to play, the most direct route to emissions reduction is the implementation of energy efficiency and the use of renewable energy.

Energy-efficient practices and technology

International energy use is steadily increasing and many governments continue to open new coal and nuclear power stations to meet the increasing demand. We need to stop our increase in energy use. One way of achieving this is through greater energy efficiency. A lot of the energy used in homes, businesses and transport is unnecessary. Improved management and technology could reduce this significantly. If we used public rather than private transport, for example, there would be significant reductions in oil emissions. Some argue we could cut energy use by 40% just by being more efficient.

Energy 'audits' can assess where energy is being used and strategize as to what behavioral and technological changes could be implemented to effect energy savings. Government laws and codes (sanctions and standards) targeting the major CO_2 producers (big businesses and the wealthy classes) could also encourage and enforce energy saving. This should include codes for appliances, buildings, vehicles, industrial processes, as well as efficient public transport. In some instances, this might entail restructuring the plan and design of cities and towns. [11]

Regulate patents and profits to make energy-efficient technology available

Government regulations should put social and environmental needs before profits, and insist that energy-efficient and renewable-energy technologies are made available and affordable.

There are numerous energy-efficient and zero-emission technologies that already exist but have been kept off the market – perhaps because they threaten the billion-dollar profits of the oil, motor and coal industries. These industries own many of the patents for these technologies. In some cases the technologies have been financially unviable, because they compete against 'cheap' fossil fuels. If mechanisms are developed to 'level the playing fields' (by factoring in the real, externalized costs of fossil fuels, or by banning or limiting their use) more of these green technologies will find their way onto the market. See 'Inspiring Examples: Groundbreaking Technology,' page 78.

38

Fact Sheet 2: The Solutions

We cannot rely on the market alone to bring these inventions to production. Even in cases where they are not suppressed, companies may bring them out slowly and incrementally, in order to maximize the profits on each stage of sales. For example, rather than bringing out the best version first, they may release models incrementally with 25%, then 50% then 75% efficiency. Many of the profits in the car industry rely on replacing and repairing parts. If they make a very efficient car, (e.g. an electric car) they would lose out on this income. The urgency for action requires that the most effective inventions are put into use at a rapid rate, driven by the needs of the environment, rather than purely by business or consumer demands. Government regulations and incentives will have to ensure this happens.

Move towards renewable energy as the *primary* source of energy

'The reserves of renewable energy that are technically accessible globally are large enough to provide about six times more power than the world currently consumes – forever.' (Dr Nitsh et al, cited in Greenpeace/EREC, 2007)

Renewable energy is energy produced from sources that naturally replenish themselves, and are therefore indefinitely sustainable. These include solar, wind, hydropower, biomass, geo-thermal, ocean energy and hydrogen. There are numerous technologies that have been and will be developed to harness the power of these renewable resources for energy production. These need to be developed and applied in a way that is of social benefit and environmentally friendly, so as not to perpetuate environmental and social crises.

Currently, renewable energy supplies 13% of the world's primary energy. Eighty percent is from fossil fuels, and seven percent from nuclear power. However, only about 1.5% of our *electricity* ('power') is generated from RE sources. [12]

We need to increase our electricity-generation from renewable sources until renewables become our *dominant* source of energy. Experts around the world have shown that this can be done. For example, *Global Energy [R]evolution*, a report and blueprint produced by the European Renewable Energy Council and Greenpeace (2007) illustrates how we can reduce the 1990 levels of CO_2 by 50% by 2050 by using energy efficiency and renewable energy practices. Their proposals factor in the growing world demands for energy and the shutting down of existing nuclear power stations.

This report is relatively conservative, and does not take into account advances in renewable technology and practice. There are ongoing breakthroughs in renewable energy technology (see 'Inspiring Examples: Groundbreaking Technology,' page 78). It is likely that as there is more research into and mass production of renewable technologies, efficiency will increase and production costs drop.

The Fire Dogs of Climate Change

Renewables are cheaper and produce more jobs

British economist, Sir Nicholas Stern, argues that acting now to reduce climate change would cost about 1% of world GDP (less than two thirds of global military spending) and that not acting would cost between 5%–20% of world GDP.

In the short term, oil and coal are relatively 'cheap' because the social, health and environmental costs are 'externalized' and not factored into its price. As I write, the price of oil is rocketing. Nuclear power is extremely expensive, even before you factor in externalized costs, and it relies on government subsidies. Renewable technologies vary enormously in form and cost. Some of them are viable only in limited circumstances, others are already cheaper than fossil fuels. Mostly they require an initial capital investment, but in the medium term they cover costs, and in the long term are more cost-effective than fossil fuels (*way* more cost effective – if you factor in the costs of fossil fuel damage). Renewable technologies also create far more jobs than coal and nuclear. For example an Earthlife Africa study (2007) found the following in relation to jobs created per Megawatt produced: Solar photovoltaic (PV) technology creates approximately 35.4; wind, 4.8; conventional nuclear, 0.5; pebble bed nuclear, 1.3; and coal between 0.7 and 3.

International studies also show that renewable energy creates more jobs per megawatt (MW) of power installed, per unit of energy produced, and per dollar of investment, than the fossil fuel energy-based sector (Environmental and Energy Study Institute, 2007).

Discourage and phase out fossil fuel; encourage and phase in renewable energy

There needs to be a moratorium on the development of coal and nuclear plants. Then we need to phase them out, increase our energy efficiency and phase in renewable energy sources. We need to research, subsidize and support environmentally-friendly renewable energy with the aim of it becoming our *primary* source of energy.

Pressure to change

Necessary changes will only be made fast enough with large-scale incentives, regulation, and enforcement of changes. This must be implemented by national governments, and enforced internationally. Where governments fail to take sufficient action, other organizations (e.g. provincial and local governments; civic, trade union, environmental and scientific organizations) should take on this role.

Fact Sheet 2: The Solutions

Now or never?

The planet's resources are finite. Supplies of fossil and nuclear fuels are limited. It is just a question of time before we have to switch over completely to renewable sources. If we switch over soon, we can prevent major environmental and social disasters.

There is no choice about *whether* to switch over to renewable energy. The question is when: now or later?

'The world lacks neither the financial resources nor technological capabilities to act. If we fail to prevent climate change it will be because we were unable to foster the political will to cooperate.' (UNDP, Human Development Report, 2007)

Running at the Enemy

I am sitting on the bathmat, my back against the wall. Bowen is in the bath. Wuppertal, is curled on a pile of his clothes on the floor. Our tummies are full and our lives are sweet, but Bowen and I feel as if there are heavy weights pressing down upon us.

'Our Emperor, George,' sighs Bowen, referring to Bush who is dropping more bombs on the people and land of the Middle East.

'All this war,' he says, a deep sadness in his voice. 'I just can't understand it.'

'The oil industry and the arms industry are making a killing,' I say.

We do not smile at my pun.

In a nearby *vlei* [1] there are thousands of tons of dead fish and eels. The waste pumped into it by neighbouring factories has caused an algal bloom that suffocated all life in the water. The bulldozers down the road are crushing the *fynbos* dunes to build ugly seven-floor holiday apartments.

Sometimes it seems as if there is a war on Life itself. Powerful governments and companies have made themselves enemies of the world. The casualties are not only people, but all the life that dances between the earth and the sky: 30,000 species of creatures and plants are becoming extinct every year. Not just getting sick or going away on holiday, but wiped off the surface of the Earth.

All the diamonds and dollars will never be able to buy this life back.

'But, you know, Bowen,' I say, suddenly struck by an obvious realization, 'we may as well believe we *can* change things. If we don't then there will be no change. And if we do then we will be motivated to act. And it's our action that is going to make the difference. The truth is, another world *is* possible – but we have to make it come true.'

Wuppertal adds her emphasis by flicking her tail repeatedly as I speak. Bowen's eyes are closed and he may have gone to sleep, so I address my speech to her: 'When I've had people at my side who are saying: "No – we won't take this anymore!" then I have felt our collective power to run at the enemy. Sometimes one creature at my side has been enough – especially when this creature was a dog.'

Wupsie glares at me disdainfully.

The dog was my Ridgeback, Riska, and we were walking by the sea, heading for the cliffs at the far side of the beach. The sand was firm under my bare feet and the sky was a clear light blue. Riska cantered at my side, her mouth wide in a tongue-wet smile, her tail a whirr of delight. Turquoise waves flattened themselves into shallow white waters. The wet sand reflected scattered patterns of bright sunlight.

The sprawling juicy-leafed plants hugged the curves of dunes that lined the top end of the beach. My skin hummed softly as it drank in the warm sun. I breathed the sea-spiced air deep into my lungs and let out a peaceful sigh.

When we were walking back along the edge of the sea, I noticed a man standing tall and white on the dunes. I quickened my pace. He moved towards us, then pulled his black shorts down below his knees. He leered at us as he touched himself. My body cringed at the invasion. Birds called overhead but no other people were in sight. The primal fear of rape surfaced to my skin, making it burn red. I wrapped a scarf around my head so that I blocked my view of him and walked briskly on.

Riska, however, did look at him. She didn't like him, and she didn't like what he was doing. She turned to face him and barked. She started walking, then running towards him. Her barks were loud and clear: 'No. We won't take this anymore!'

I threw the scarf off my face and turned and ran with her. We were charging at him together. He started to run away, awkwardly at first with his shorts around his ankles, then faster when he pulled them up. He ran and we chased him down and up and over the dunes to the road. We caught sight of the tail end of his *bakkie* [2] with its backfiring exhaust-pipe as it roared away into the distance.

I patted Riska's head and she chewed my ankle gently. Then we continued our walk, back over the dunes and along the sun-sparkled wet sand.

I scratch Wupsie under the chin and she forgives me my dog-talk. I tell her: 'Sometimes you need a bigger group of people if you want to run at the enemy.'

In the 1980s we ran with an energetic bounce in our step. It is called the *toyi-toyi*. I can still hear an old song fresh in my ears:

Inzima le ndlela, inameva, iyahlaba; sizozabalaza (It is difficult this road, it has thorns, it stabs; we will struggle on). We are singing and dancing. We are angry and jubilant, we are mourning and celebrating. *Sizozabalaza.* The dust is under our feet. We are stamping our feet and marching and dancing and running. We are safe. We are surrounded by enemies. We are unarmed. We are dangerous.

The Fire Dogs of Climate Change

I run next to a young woman. I have never met her. She is holding my hand. I have always known her. We are singing, we are shouting. We are strong. We are young and old. We pass people standing in their front yards. We sing: *Molweni. Molweni. Ninjani? Ninjani? (Hello. Hello. How are you? How are you?)* We are surrounded by friends. I came with one friend but I lost her. For a moment I felt alone, then I was washed into our pulsing angry joyous strong song, moved along to the graveyard to bury our dead, surrounded by the armoured trucks of the people who killed them. Surrounded by their guns and steel, we are running and dancing and singing.

Suddenly we hear shots. The sounds of the bullets punch through my heart and the blood runs rushing, rushing through my body and brain. I look for my friend. My first friend, my new friends. We are running. I am afraid. I am running as fast as my heartbeat.

I think we are running to safety to escape the bullets, then I see ahead the tear gas and hear again the shots and I realize that we are running at the enemy, running to fight.

We did not win the fight that day; nor the next. Many more were buried. But we did in the end beat that enemy. We slaughtered the fat pink beast of Apartheid, and offered it to our ancestors. But the clever creature, Capitalism, stayed alive and fed off the fat.

I have sometimes been lucky to be beside those who are brave and strong enough to run at the enemy. Workers fighting: for their lives, for fair wages, for the right to vote, for another world often named socialism.

After some years of running I was forced to lie down for a long time. My daily fight was with an illness inside my own body. The only running I could do was running a bath – and even this was difficult. Most of the time I lay very still on a bed in a dark room, and ached and breathed.

Occasionally I felt strong enough to lie on the grass under the shade of our orange tree and listen to the humming of the bees, which buzzed around my head, as if looking for my pollen. I would watch the hoopoes, with their wild black and orange headdresses, stabbing long beaks into the soil; and the black fork-tailed drongoes, swooping down in crescents from the branches. Beneath me was the solid earth and grass; above, the empty deep sky. Between the two, life danced and sang around me.

I was cared for by a man and two dogs who loved me. The dogs would offer soft-mouthed greetings and happy wriggles. The man would hold me in his arms when I cried. Having them at my side gave me strength to keep on fighting. I went to a clinic where I drank water for fifteen days. It slowed me right down to a feeble crawl. My mother came and carried me outside into the sun.

The water washed through and new life moved into me. I ate only fresh raw fruit and vegetables and my body began to taste its own strength. For the first time in two years I went for a ten-minute walk. The blood pumped through my limbs as I moved one leg forward then the other, my arms swinging at my sides. I could walk. I was striding! My smile stretched wide across my face and spread through my pulsing blood, each step confidently feeling the earth beneath my feet. Not running, but walking.

Right now, sitting in our bathroom, we are not running or walking, but talking; looking for ways to push off the weights of hopelessness and powerlessness we sometimes feel.

'We need to remember,' I say to Bowen and Wupsie, 'that people all over are fighting.'

There are many fire dogs, guardians and watchdogs of the community and Earth. I spoke to one this morning.

His name is Desmond D'sa and he is the chair of the South Durban Community Environmental Alliance.

'We are watchdogs,' Desmond told me. 'That's our job.'

He lives in a working-class area on the fence line – next to the Engen oil refinery. The multinational oil companies are the biggest polluters in the South Durban area. They burn toxic gas and chemicals into the air and leak them into the ground and water. Among the people who live in this area, there are very high incidences of asthma, cancer and leukemia.

I asked him how he keeps going and he said: 'I wake up every morning, feeling strong, and I say to myself: "I am going out there into battle."'

'Sometimes there is a strong smell like rotten eggs in the air,' said Desmond. 'Kids are hit worst by the pollution. I was once called to a primary school by the principal. I watched children dropping down to the ground in front of me. Engen didn't want to pay to get the kids to the nearby hospital; they said the kids were making it up. We were so angry with them. We told them: "If one of these kids dies you are going to have all hell to pay. We will come in thousands and blockade your gates. We will close your refinery down." So they agreed to pay.

'We have blockaded their refineries before, so they know we can do it. In 2001, the Construction, Engineering and Industrial Workers Union organized a big strike at Engen. Six thousand people came and stood outside Engen's gates.

The Fire Dogs of Climate Change

We marched up and down in the streets of the community. We sang songs and carried banners. We shut down that plant and they were forced to listen to the workers.'

Every day Desmond goes to meetings of people organizing and fighting for Environmental Justice. 'And we have had many victories,' he said. 'We have made Shell replace their old and leaky pipelines. We have pushed government to make and enforce pollution laws. We have education projects for the youth.

'It is not just the pollution that must end,' continued Desmond. 'Why should these companies hold so much power in their own hands? Our only hope is if we keep fighting. Sometimes we lose in the short term but in the long term we will win. We are riding the dragon.'

'We may as well believe we can do it,' I say again to Bowen.

He opens his eyes and nods. Wupsie stretches out her claws in agreement. Then we all sit quietly in the bathroom. I remember a quote by Arundhati Roy: *'Another world is not only possible, she is on her way. On a quiet day, I can hear her breathing.'* [3]

Inspiring Examples:
Action

Would you like to take action: whether it is turning off a light, chaining yourself to a member of parliament, or launching a national campaign? All over the world there are fire dogs running at the enemy. There are many different types of dogs, running alone and in packs; and they are barking at a whole lot of dangers. See if there are any packs that you would like to run with...

'Shut down the coal port; green jobs now!'

Anna Rose gives us a taste of one of the actions taking place around the world. On 13 July 2008, she writes about the third day of a Climate Camp: 'Right now I'm feeling so excited and happy about what happened today in Newcastle, my hometown in Australia. Around 1200 people today took direct action to stop the disastrous environmental impact of the world's biggest coal port in Newcastle. The spirited and colourful protest was made up of a diverse mix of people including families, coal workers and activists ... even some zombies, clowns, and radical cheerleaders. Many people made it onto the rail line – through or under the fence – and coal transport in Newcastle was shut down for the entire day. No coal trains got through. Organisers estimated that we cost the coal industry about 1.3 million in lost revenue. There were 50 arrests, and most arrestees were let off with a $400 fine, which we will all fund-raise for in the next few months. The mood was inspiring and strong, with the crowd chanting, "shut down the coal port; green jobs now."

'Coal exports are Australia's biggest contribution to climate change and the greenhouse pollution from our coal exports exceeds all of our domestic pollution combined.' (www.itsgettinghotinhere.org)

The Fire Dogs of Climate Change

It's Getting Hot in Here

It's Getting Hot in Here is a collection of voices from the student and youth leaders of the global movement to stop global warming. Originally created by youth leaders to allow youth to report from the International Climate Negotiations in Montreal, It's Getting Hot in Here has since grown into a global online community, with over 100 writers from countries around the world.

(www.itsgettinghotinhere.org)

Climate Action Network (CAN)

CAN is a worldwide network of over 400 NGOs from around the world working to promote government and individual action to keep global warming as far below 2°C as possible.

CAN members work to achieve this goal through the coordination of information exchange and NGO strategy on international, regional and national climate issues. There are regional offices in Africa, Central and Eastern Europe, Europe, Latin America, North America, South Asia, and Southeast Asia. The NGOs vary a lot, some are wild and sharp-toothed, and others are well-groomed and trained to roll-over. Check out their website to find NGOs in your country.

(www.climatenetwork.org)

United Nations and the International Panel on Climate Change

These dogs have been very busy when it comes to climate change. The United Nations Framework on Climate Change (UNFCC) is the guiding framework for government policy and action (and inaction).

The UN helped get the whole International Panel of Climate Change (IPCC) pack of scientists together and they all barked and barked. And barked. The IPCC furrow their furry brows, and lie down to think a lot, and then get up and keep on barking like those toys with long-life batteries. Some people say that although they bark loudly, the UN and IPCC don't have enough bite: they are kept on a leash by governments, and trained to listen to the whistle of big business. But they seem to break free whenever they can and follow their clever doggy hearts. They do their best to shepherd world governments in the right direction, but sometimes this is as difficult as herding cats.

(http://unfccc.int/; www.ipcc.ch)

Inspiring Examples: Action

Al Gore gets bolder (and wiser)

'Today I challenge our nation to commit to producing 100 percent of our electricity from renewable energy and truly clean carbon-free sources within ten years.' – *Al Gore (2008)*

While most governments are taking baby steps (and expecting loud claps), and many mainstream environmental organizations are pussy-footing about (claiming to be 'realistic'), Al Gore reminds us that we walked on the moon. He urges us to once again take that 'giant leap for mankind.'

Al Gore may have been an obstacle to progress when he represented the US in the first climate change negotiations, but he has not dedicated his retirement from public office to sitting back and growing old. Instead he grows bolder. His movie, *An Inconvenient Truth*, swept across the world like wildfire, raising awareness about the problem of climate change. And now he is telling the truth about the solution. I quote a number of extracts from his groundbreaking (earth-healing) speech in July 2008:

'This goal (100% RE electricity in ten years) is achievable, affordable and transformative,' he states. 'Of course there are those who will tell us this can't be done. Some of the voices we hear are the defenders of the status quo – the ones with a vested interest in perpetuating the current system, no matter how high a price the rest of us will have to pay.'

Gore argues that the over-reliance on fossil fuels in the US is at the core of the economic, environmental and national security crises.

'I for one do not believe our country can withstand 10 more years of the status quo. Our families cannot stand 10 more years of gas price increases. Our workers cannot stand 10 more years of job losses and outsourcing of factories. Our economy cannot stand 10 more years of sending $2 billion every 24 hours to foreign countries for oil. And our soldiers and their families cannot take another 10 years of repeated troop deployments to dangerous regions that just happen to have large oil supplies.'

'We're borrowing money from China to buy oil from the Persian Gulf to burn it in ways that destroy the planet. Every bit of that's got to change.'

The solution he proposes: ditch fossil fuels; switch to renewables. Now.

Gore urges Americans to set a leading example for the world, and encourages us all to join the 'WE campaign'. This campaign is 'committed to changing not just light bulbs, but laws.'

(www.wecansolveit.org) [1]

The Fire Dogs of Climate Change

The United Nations Environment Programme (UNEP)

UN and UNEP produce very good reports and publications, and have great climate change education programmes and youth programmes. Have a look at the UN Environment Programme Climate Neutral Network and the UNEP publication, Kick the Habit: A UN Guide to Climate Neutrality.' [2]

Around the world, UNEP-initiated World Environment Day events highlighted resources and initiatives that promote low carbon economies and life-styles, such as improved energy efficiency, alternative energy sources, forest conservation and eco-friendly consumption. [3]

UNEP and Primary Project initiated Clean Up the World – a community based environmental campaign that inspires and empowers communities from every corner of the globe to clean up, fix up and conserve their local environment. Since 1993, Clean Up the World has inspired an estimated 35 million volunteers in 120 countries each year to take action. They are now promoting the theme: 'Clean up our Climate.'

'Climate Action' is an international communications platform, produced by Sustainable Development International in partnership with UNEP. It is designed to assist the private and public sectors towards carbon neutrality, as well as providing practical actions to reduce our global carbon footprint.

(www.unep.org; www.cleanuptheworld.org;
info@climateactionprogramme.org)

Greenpeace International

You should all have heard of Greenpeace because they bark very loudly and clearly. They are a wild but well-disciplined bunch of dogs that know how to sit and stay. They warn us that we must keep as far as possible below 450ppm CO_2e. I mention some of their successful actions later on. Go to their website, and get their monthly newsletters to see how you can get involved in their climate change activities.

(www.greenpeace.org/international)

Friends of the Earth International (FOEI)

FOEI is the world's largest grassroots environmental network. They are barking (and biting too, sometimes) for environmental justice. They are brave, hairy dogs working with communities to fight multinational companies that are trashing the planet. They state: 'Friends of the Earth is calling for urgent action to stop humans intensifying climate change, the biggest environmental threat to the planet. We are demanding strong national emissions reductions targets, and have initiated lawsuits against the world's worst polluters, including major oil corporations, the US government and financial institutions. We are challenging a number of big oil projects around the world that will accelerate climate change. We have also joined forces with climate-affected communities to build a global movement that addresses social and economic equity between and within countries.'

Some of their actions are listed later in this section. Find the FOEI affiliate nearest to you. There is also a 'Young Friends of the Earth.' You may have seem some of these fire puppies dressed up as penguins, and marching through the streets of Berlin, with the aim of speeding up the negotiations in Bali.

(www.foei.org)

Campaign against Climate Change, and Stop Climate Chaos (UK)

George Monbiot (who in my opinion has one of the sharpest brains on the hottest subjects) is president of this UK organization, Campaign against Climate Change. (Have a look at his book, *Heat*, 2006.) The Campaign actions are focused, organized and colourful. The Campaign against Climate Change is pioneering a global climate change campaign, and are part of the broad coalition: Stop Climate Chaos.

(www.campaigncc.org; www.globalclimatecampaign.org;
www.stopclimatechaos.org)

Trade unions and worker action

Trade Unions are represented in the UNFCC and have been involved in climate change agreements. International trade union bodies support the IPCC calls for substantial reduction targets and appeal to workers to help save our planet. [4]

Many trade unions are part of broader climate change networks, and have internal climate change programmes in place. A conscious, organized working class has the power to make political and economic changes. They have the direct experience of the damaging effects of fossil fuels, and an understanding of capitalism's drive for profits (at the expense of people and planet). They work in the businesses that are responsible for the greatest emissions and have the power to disrupt or restructure this work. One of the trade union responses to climate change has been workplace greening initiatives. [5] But workers also have their livelihoods to protect; and jobs that are ended with the fossil fuel industry, must (and can) be replaced by 'green jobs.' Organizations and actions that ignore the working class may make enemies of potential allies. A problem with some of the sporadic and 'direct action' activities, is that they

The Fire Dogs of Climate Change

fail to build relationships with the workers. Green activist groups may perceive themselves as 'radical' and workers as 'conservative', when the workers act to protect their jobs, or express anger at not being consulted. In the UK, there is a 'Workers climate action' initiative aimed at building a bridge between labour and environmental movement, and to try and find common interests and solidarity. The International Labour Foundation for Sustainable Development aims to integrate sustainability issues into the labour movement.

The Trade Union Sustainable Development Unit maintains country-by-country profiles on the progress made by governments along sustainable development indicators, including climate change and energy. [6]

(www.unions.org; www.sustainlabour.org) [7]

Women who run with the dogs

Women have been active in every level of climate change action. They are often the worst affected by climate changes. Women in poor rural areas, for example, bear the brunt of the stresses of drought on domestic life and agriculture. They have to walk longer distances to collect water, and make meals from diminishing crops or gathered food. There are specific initiatives dealing with gender and climate change.

(www.gendercc.net; www.siyanda.org)

350 organization

350 is an international organization run mostly by youth, pressurizing governments to commit to bringing CO_2 levels down to below 350 parts per million. They love to run around in open spaces and howl across long distances.

(www.350.org)

'No Compromise in the Defence of Mother Earth!'

Earth first! is a radical 'direct action' international environmental movement. Their emblem is a green fist and their slogan is 'No Compromise in the Defence of Mother Earth!'

They introduce themselves by asking: 'Are you tired of namby pamby environmental groups?' and state that 'while there is broad diversity within Earth First! from animal rights vegans to wilderness hunting guides... from whiskey-drinking backwoods riffraff to thoughtful philosophers... there is agreement on one thing: the need for action!'

Inspiring Examples: Action

They are acting on the most urgent and acute wounds to the Earth, i.e. 'stopping the bleeding.' Earthfirst! is a non-hierarchical movement, with small groups that make decisions by consensus. 'Leaders' are those who are taking effective action, not those who have organizational control. Earthfirst! philosophy is rooted in Deep Ecology: driven by spirit and heart as much as by a political fight against injustice.

If you are ready to chain yourself to a tree, or blockade a coal station, you should run with these wolves. They love to howl and sing in wild places. Movements such as Rising Tide, Plane Stupid, and Climate Camps have a similar non-hierarchical organization and 'direct action' approach.

(www.earthfirst.org)

Rising tide

Rising Tide is an international, diverse group of individuals and organizations that are committed to taking action on the root causes of climate change. They argue that climate change is a result of social injustice and that the Kyoto protocol is bound to fail because it depends on flawed market mechanisms. Rising Tide's aim is to cut fossil fuel use and move to a non-carbon society as soon as possible. without making the poor and the working classes bear the burden.

(www.risingtide.org.uk)

Plane stupid (UK) and an inflatable elephant

Emissions from aircraft are especially problematic because of the height at which they are emitted, and the particularly noxious mix of gases; making them 2.7 times more damaging than the effect of their carbon dioxide alone. UK airports are rapidly expanding yet this sector is not considered by any binding international treaties such as the Kyoto protocol. Plane Stupid is pretty clever at finding creative ways to get their message across. For example Plane Stupid Scotland managed to sneak a 5-metre high giant inflatable elephant into an aviation conference at the Edinburgh Caledonian Hilton – with a massive banner stating, Aviation is the elephant in the room.'

(www.planestupid.com)

Camp for Climate Action

In 2006 about 500 people camped at Drax power station in the UK, with the intention of shutting it down. This inspired annual climate camps where people demonstrate, and experiment with sustainable living. The idea has taken off in Australia and is spreading

The Fire Dogs of Climate Change

to other countries. Monthly planning meetings are held, and decisions are made by consensus. There is no central coordination and working groups are established to do different things. A series of action days are planned in these forums – such as 'April Biofools day' and Network For Climate Action. [8]

(www.climatecamp.org.uk) [8]

Many Strong Voices – from the islands

Many Strong Voices (MSV) is a climate change mitigation and adaptation programme, linking coastal communities who live in the Arctic and Small Island Developing States (SIDS). These communities are some of the hardest hit by climate change (melting ice and rising seas), yet they also have a history of self-sufficiency and adaptation . These fire dogs have a lot to teach us, as their paws are already getting wet. La Réunion island (not an MSV member) has started a 'green energy revolution,' and plans to become carbon neutral by 2030.

(www.manystrongvoices.org; www.gerri.fr) [9]

The 11th hour

The 11th Hour, a movie produced and narrated by Leonardo Di Caprio, brings the message home to youth; and his website provides ideas on actions we can take.

(www.11thhouraction.com)

Celsias: practical projects

Celsias claims to be the 'the world's leading action-based climate change website'. But the actions that they are referring to are not the political actions discussed above – but practical actions and projects that directly reduce emissions. Their slogan is: 'Climate change is not a spectator sport.' They urge you to 'learn, commit and do.'

Celsias lists a number of emission-reducing ideas, projects, and technology – inviting you to support them, and to post your own. These include ideas on how to reduce your carbon footprint. They have a vibrant group of journalists from around the world sniffing out breaking environmental news, and inspiring initiatives by individuals and businesses. They provide a crucial reminder to us that change is not just about fighting against what we don't want, but creating a practical alternative.

(www.celsias.com)

Inspiring Examples: Action

Green Thing

Green Thing is a new community that makes it easy and enjoyable to be a bit greener. Every month you'll get a different Green Thing to do. All you have to do is do it. Over 123,000 people from 168 countries are participating, and at least 3,000 tonnes of CO_2 have been saved so far.

(www.dothegreenthing.com)

Fight global warming with your fork

The US Center for Food Safety and the Corner Stone Campaign has launched the Cool Foods Campaign to educate people about the everyday changes they can make to fight global warming with their fork. They estimate that about 25% of US greenhouse gas emissions are from the current industrial food system (including agriculture, transport, processing etc.).

The Campaign involves businesses, organizations, farms, schools and individuals who have signed the Cool Foods pledge and are changing the way they eat. Based on scientific analysis of what food-related processes cause the most GHGs, the Cool Foods Campaign advocates lower 'FoodPrint' alternatives; including organic, local, whole foods, less processed and packaged, grass-fed, free-range or wild-caught meat and seafood from sustainable fisheries.

(www.coolfoodscampaign.org)

Science activists make forests 'valuable'

The Global Canopy Programme is an Oxford-based alliance of science organizations around the world specialized in canopy science. Their mission is to help make forests worth more alive than dead by using science to inform policy and finance mechanisms. This year they helped launch Canopy Capital, an investment mechanism for standing forests. While this is not traditional activism, it is nevertheless a contribution to the battle of ideas about how the global economy values (or fails to value) forests, forest communities and half of the species of life on earth that live there.

'Canopy Capital' was established in 2007 to drive capital to the rainforest canopy. Canopy Capital has created an investment template for first-movers in an emerging market for Ecosystem Services. These include rainfall generation, moderation of extreme weather, carbon storage and biodiversity maintenance. These services benefit humanity at local to global scales. If they are lost, there will be severe impacts on food, energy, and environmental security. Putting a price on these services

The Fire Dogs of Climate Change

is like taking out an insurance policy to maintain our life support system and has the potential to generate billions of dollars for forest-owning nations.'

(www.globalcanopy.org; www.canopycapital.co.uk)

Though it is crucial to protect our natural resources, personally I'm not convinced that this is the best way to go about it. Forests should be unconditionally protected by governments, and not be subject to the whims of the market (or the carbon-offset market). Nevertheless they *are* currently in the hands of the market and this approach may contribute to protecting them in the short term. For what is also necessary to protect our forests, see the statement, 'Protecting the world's forests needs more than just money' – signed by organizations from around the Earth.

(www.foei.org)

Below are some of the forest protection victories, which have come about as a result of political struggle.

Forest victories

2006: A decree by President Lula of Brazil calls for a 6.4 million hectare (around 16 million acres) conservation area protecting an area of the Amazon from land-grabbers, cattle ranchers and loggers.

2006: McDonald's agrees to stop selling chicken fed on soya grown in newly deforested areas of the Amazon rainforest. Other food companies and supermarkets (e.g. Marks & Spencer, Sainsbury's, ASDA and Waitrose) sign up to a zero deforestation policy. Multinational soya companies such as Cargill agree to a two-year moratorium on buying soya from newly deforested areas.

2007: Greenpeace and other environmental groups get 1.5 million signatures of support and push through Argentina's first federal forest protection law.

2008: Soya moratorium in the Amazon is extended for another year.

(www.greenpeace.org) [10]

Here are a few more examples of inspiring social action ...

Poor fight agri-business

All over the world, groups of working class and rural poor are taking on the giant monoculture agribusinesses (growing biofuels and GM crops) that are pushing them off their land and making food too expensive to buy. In every country you will find reports of

56

Inspiring Examples: Action

Figure 1. Mapping the global variation in CO_2 emissions *(United Nations Development Programme (UNDP), 2007)*

The Fire Dogs of Climate Change

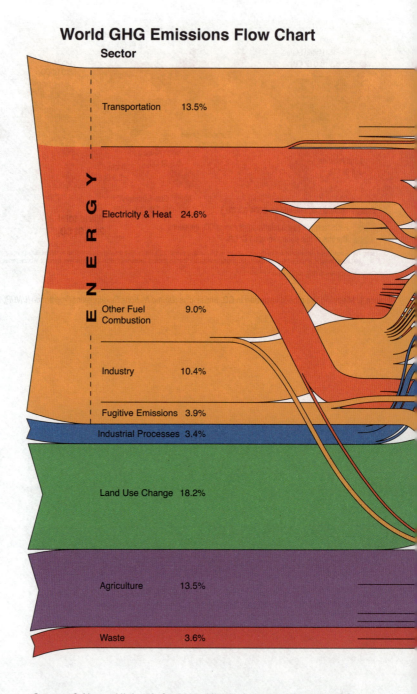

Sources & Notes: All data is for 2000. All calculations are based on CO_2 equivalents, using 100-year global warming potentials from the IPCC (1996), based on a total global estimate of 41,755 Mt CO_2 equivalent. Land use change includes both emissions and absorptions.

Inspiring Examples: Action

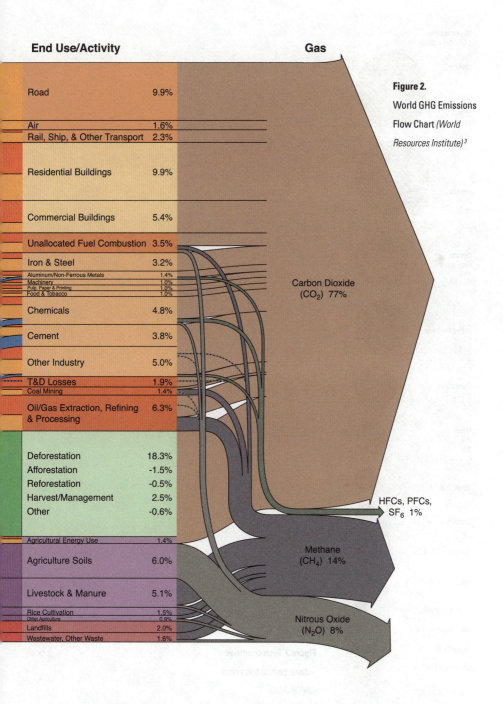

Figure 2.
World GHG Emissions Flow Chart *(World Resources Institute)*[3]

The Fire Dogs of Climate Change

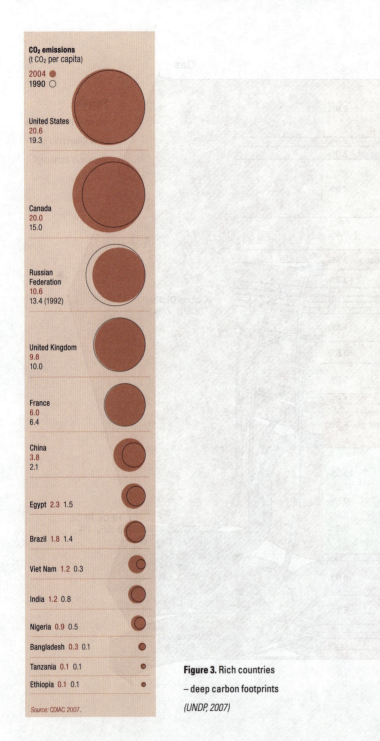

Figure 3. Rich countries
– deep carbon footprints
(UNDP, 2007)

Inspiring Examples: Action

people struggling against the businesses who are making profits at their expense.

In 2007, in Argentina and Italy, protesters went without tomatoes and pasta in food boycotts against rising prices. 25,000 people marched on Delhi in protest against land grabbing for big business agriculture. In Mexico, tens of thousands protested against the fast rising price of maize. As biofuels push up the price of staple foods there have been increasing food riots and protests. Latin American, African and Asian social movements, unions and NGOs have made declarations highlighting their concern over the ways in which agrofuel production is taking land from subsistence farmers and destroying national food sovereignty. Latin American NGOs stated that 'Land must be used to feed people, not cars'.

Social organizations, movements and pastorals in Brazil declared 'Agroenergy policies cannot continue to be determined by market logic, and by the interests of oil companies, car industry and agrobusiness... We demand the end to deforestation... We affirm the sovereignty of traditional peoples and communities over their territories' (1st National Popular Conference on Agroenergy, 2007).

(www.foei.org; www.biofuelwatch.org.uk) [11]

New Zealand power plant cancelled

In 2007, the New Zealand government announced cancellation of the proposed coal-burning power plant Marsden B. Greenpeace and local activists had mounted a four-year struggle which involved a nine-day occupation, high court challenges, protest marches, a record numbers of public submissions, Surfers Against Sulphur, public meetings, and a pirate radio station.

(www.greenpeace.org) [12]

'Oil is the blood of Mother Earth'

> 'Oil is the blood of Mother Earth. It belongs to the ground, where it sustains the world below. Up here it only causes violence and death.' – Luis Cabellero, vice president of the Traditional U'wa Authority, Columbia.

The 7,500 U'wa people, aided by Colombian social movements, environmental organizations (including Friends of the Earth Colombia) and campaign groups around the world, succeeded in their decade-long campaign to stop the exploitation of their land.

'Their stabbing was felt some feet below the Earth's surface and she cried. She asked us to defend her and to tell the world what she was feeling. We did that and we will continue to do that.' – The Traditional U'wa Authority.

(www.foei.org)

The Fire Dogs of Climate Change

Friends of the Earth sues the Enemies of the Earth

Friends of the Earth International hosts the Climate Justice Programme, launched in 2003. Dozens of organizations and lawyers have collaborated to support law enforcement around the world to combat climate change and associated human rights abuses.

Legal challenges are underway in the United States against the Bush administration's export credit bodies, for not taking climate change into account when providing financial support for fossil fuel projects; and against its Environmental Protection Agency for rejecting its power under the Clean Air Act to regulate global warming emissions. A 2004 Friends of the Earth International publication (about Exxon Mobil) has shown how we can establish liability of major corporations for global warming.

Friends of the Earth and Greenpeace members involved in a US law-suit include a North Carolina couple who fear their retirement property will be lost to storm surges, erosion and the rising sea level; maple syrup producers in Vermont who believe their business will be ruined as maple trees disappear from the area; and a marine biologist whose life's work is in jeopardy because the coral reefs he has spent a lifetime studying and enjoying are disappearing at an alarming rate due to bleaching from rising ocean temperatures.

(www.climatelaw.org/media)

Solar electric rock trip across India

A bunch of young Indians and Americans set out on a six-week 'climate solutions road trip' across India in Reva electric cars with solar panels (see Reva under 'Inspiring Examples: Groundbreaking Technology'). They travelled with a rock band – which played solar-charged musical instruments. The trip ran from Chennai to Delhi, stopping along the way, making music, educating, and discussing climate change solutions. They showcased cool technology and collected examples of practical solutions from across India. Go to their website for reports on the adventures of this pack of travelling dogs.

(www.indiaclimatesolutions.com).

Climate Bike Riders: Naked and Clothed

World Naked Bike Ride does just what it sounds like: people ride naked on bikes, in different places around the world. They promote the use of bicycles and raise awareness about climate change, and celebrate their bodies. Some paint themselves with

Inspiring Examples: Action

messages or designs, some dress up, others are stark naked; they shout on loud-hailers, chant slogans (e.g. 'burn bras not oil') and make people laugh and think.

Brita Climate Ride in the US, is one of a series of (clothed) climate rides to raise money and awareness about climate change. Starting on September 20th 2008, one hundred Climate Riders pedalled 320 miles from New York City to the nation's Capitol in Washington D.C. Along the way, expert speakers educated and inspired Climate Riders about the science, the policies and the solutions to the climate crisis. Climate Riders raised $2250 each in order to participate; all proceeds went to Focus the Nation and Clean Air – Cool Planet.

(www.globalclimatechangeaction.org/wnbr; www.ClimateRide.org)

Fun plays and educational projects

There are a number of other fun and creative educational initiatives, many of them aimed at youth. The Otesha Project is an outreach programme that started in Canada and spread to the UK. Young volunteers on bicycles perform a travelling play about making positive choices related to energy use.

TRAPESE (Taking Radical Action through Popular Education and Sustainable Every-thing) has toured around the UK, using games, films and campaign-planning as part of their climate change workshops.

Save Our World sponsored a climate change road show that subsequently took off in UK schools.[13] Kyla Davis has brought the play to South Africa where she has established Well Worn Theatre, which takes educational programmes to primary schools. They raise awareness, and facilitate energy-saving actions in the school. But, she states, 'the most important aspect of the programme is that it leaves the children feeling empowered and positive' about finding solutions for a daunting set of issues.

The 'high-energy, action-adventure' play is the 'story of Eco, a passionate young eco-warrior and Vusi, a lazy ten year old boy. Together they go on a mission to save the Bengal tigers from the rising seas and stop Global Warming. Along the way, Greedy Gutman tries to lead them astray, trying to convince them to EAT! BUY! CONSUME! They resist however, and finally Vusi returns home a changed boy and enlists the help of the audience in finding ways to STOP! and SAVE ENERGY!'

Other exciting educational initiatives include: COIN (Climate Outreach and Informa-tion Network) the UNEP youth projects, UNEP children's publication, and the UNEP/UNESCO youthXchange project. The Cape Farewell Project has also run educational tours for youth, taking them all the way to the Arctic. [14]

(www.otesha.org.uk; www.trapese.org; www.wellworn.org.za)

The Fire Dogs of Climate Change

The policy makers, carbon-movers and paper-shakers

Yes, there are even fire dogs in government, and amongst the consultants, NGOs and advisers close to government. There are a lot of climate change professionals, academics and activists involved in researching and advising on policy, and although some of them promote vested interests (like nuclear energy and biofuels) there are many fire dogs amongst them, who doggedly push on, through long papers and boring meetings. They are well-groomed dogs, trained to be very patient, and they are even capable of leading the blind.

Together with the wild and barking dogs they pushed big changes in government policy over the last few years. Governments have moved from saying, 'What climate crisis?' to making substantial commitments. In most cases the targets they have set do not go as far as is necessary, so all the dogs have to keep on barking and trying to herd them in the right direction. The movers and shakers are the ones that are closest to the government's legs, and positioned for some good ankle-nipping (Climate Corgis use this method to get even the slowest of cows moving).

The best way to link up with packs of policy makers and carbon professionals is through climate list – a knowledgebase of International Climate Change Activities, provided by the International Institute for Sustainable Development (IISD) in cooperation with the UN Chief Executives Board for Coordination (CEB) Secretariat.

The Climate Change and Energy team at IISD has been exploring a wide range of perspectives and ideas on how to develop a sustainable integrated global response to climate change.

You can subscribe to IISD Reporting Services' free newsletters and lists and to Carbon Positive News. [15]

Amongst the best preened of the movers and shakers is the Pew Center on Global Climate Change. They have loads of climate change website links for national and international climate change policy and strategies. You will find detailed government climate change policies on the International Energy Agency (IEA) website: IEA also has policy data bases on energy efficiency and renewable energy. [16]

Many of these professionals are involved in carbon trading and carbon offset programmes and policy. For more critical perspectives on this subject, see The Corner House. [17] They growl at the carbon-shufflers. Oi, dogs! There is always a bit of fighting amongst them.

(www.climate-l.org; www.iisd.org/climate;
www.pewclimate.org; www.thecornerhouse.org.uk)

Inspiring Examples: Action

Governments don't have to wait to be told what to do

In some cases, where national governments are slow to take the necessary action, provinces or states have set their own goals. This has happened in a few places in the United States, and in my home province, the Western Cape in South Africa.

Although classified as a 'developing country', South Africa has the highest GHG emissions in Africa, and high per capita emissions even by international standards. In 2007, the Western Cape set significant short-term targets for renewable energy, energy efficiency and CO_2 reductions. In 2008, the South African national government made long-term commitments (including peak emissions by 2020) that set a positive precedent for other developing countries.

(www.wcapeenergy.net)

Finding your fire dogs

So how do you choose which pack to run with? Follow your nose, and see what makes your tail wag and your paws dance...

65

Dancing with Fire

'Ouch!'

I have hit my head with a tennis ball. It is dark, and I am in my back garden. In each hand I am swinging a chain with a ball attached to it. I move them around in figure-of-eight patterns above my head. My cat looks at me like I'm crazy, but when I get it right she comes to sit smugly at my feet. My left wrist is moving over and under my right in an impossible dance and the balls are spinning, miraculously not crashing into each other or me. I am elated. It feels like someone has oiled my creaky brain. I immerse myself into movement that I cannot understand, but which makes my whole being sing.

I am learning to dance with fire. When I am ready, I will replace the balls with burning wicks, and I will dance. If I get it wrong, I will get burned. It will be my trial by fire.

It has been a good year to learn to fire-dance: the Year of the Fire Dog, according to Chinese astrology. And in African cosmology (a *sangoma* [1] told me) it is the Year of the Fire Gods. [2]

'The fire-spirits are close by,' he said. 'It is the year for trials by fire.'

'Global warming,' I said. 'That's the trial by fire for humans. We must learn or we must burn.'

As I spin the balls around my body, I look up at the dark night sky, scattered with stars, and I feel as if I am learning one of the oldest dances of the universe. I sense the Earth under my feet – a spinning ball of rock dancing around a giant sphere of fire. Spheres of rock or gas are moving around each of the billions of fire-stars in the sky.

Most of these balls are dry and dead, but the one I am standing on has a humming layer of life. Protected by a swirling atmosphere, it dances with the sun. I want to take part in this dance – the dance with fire that makes life sing around the curves of the earth and through the cells of my being.

This year I have been learning about the dance with fire: the Earth's dance with the sun; humans' dance with the sun's energy; and my own dance with the fire inside of me.

I have learned by watching. I have seen that some people are more intent on burning than learning: they drop bombs on children in their pyjamas; make smoky fires of coal ripped from beneath the warm crust of the earth; and torch tangled forests.

On some quiet nights I could hear the planet gasping.

But I heard something else too: the sound of barking. The fire dogs have been feeling the warmth of the earth under their paws. Their ears have pricked up, and their noses are twitching. They howl. They run in packs and bark to protect the Earth. And as they run, they are dancing.

I have looked for ways to be part of this dance: in my heart, home and politics.

My first dance movements were tentative. I started close by. In the cold night wind I went with my man, Bowen, and our neighbours' children to the hollows and hills of the dunes across the road. The bulldozers were coming the next day to crush this narrow vein of life that pulsed with fynbos and small, shy creatures that shelter in its leaves and sandy burrows. We shone our torches on the tips of the swaying plants; and searched late into the night to find a few glowing chameleons gripping tight, their eyes closed in sleep. They crawled onto our warm hands and we carried them to an open plot that is safe, for now, from the bulldozers.

We took food to the young sea dogs, on the *Sea Shepherd* ship at the Cape Town docks, who fight to protect the seas from illegal whaling and netting. We collected donations to stock up their threadbare workshop. Bowen and I wove through the roads of Muizenberg on our bicycles, visiting friends and neighbours and tool shops. We got angle grinders, drill bits, nails and paintbrushes. Each person that gave is now linked by a thread to the ship. These scores of threads are part of the weave that supports the craft, which now dips and dives through the waves.

As well as taking small dance steps close-by, I took cyber-leaps across the worldwide web. I found many stories of fire dogs who were engaged in the dance with fire: fighting the fossil-burners and protecting life.

Even in the icy Arctic, there are fire dogs: the Gwich'in – or Caribou People. They live on the northern tip of the planet, in a place where luminous ribbons of colour glow and flow across the darkened sky. When the thick white cover on the earth melts, and the green soft grass pushes through, this cold and wild place fills with millions of birds that have travelled the curves of the earth to meet and eat and make love and make life. Big woolly creatures, with horns like gnarled wood, cross the melting snow and furry white bears slip into icy waters and chase the sleek fast seals. Giant gentle mammals swim around the globe to feed in these seas.

But all is not well in this icy Eden. Thick waves of toxins from the industries and fields of its southern neighbour, the United States of America, are washed up in the currents of the sea and air to settle into the mossy lichen and the tissues of the tiny fish. Poisons permeate the blubber of the sea mammals and the bones of the reindeer.

And the land of ice is slowly melting under the heat that has been trapped by the greenhouse gases. The seals and polar bears need deep snow and large ice floes to build the snug dens where they give birth, suckle and hibernate. As the ice floes shrink, so too do the schools of krill, which breed under the shelter of its surface. The rich and tasty krill is the basic food for the whales and many creatures of the sea. A small increase in heat may seem like just a drop in the ocean, but it causes giant ripples across the sea of Arctic life.

Even with all these pressures and threats, the Arctic National Wildlife Refuge remains one of the most complete and undisturbed ecosystems left on Earth. A safe place where snow geese can mate and wolves can run wild through tree-lined mountains and open plains.

The oil companies and their allies have been trying to drill here for oil. The area they hope to mine is the breeding ground of the caribou – which is the biological heart of the Refuge. The Gwich'in call it 'the sacred place where life begins'. The oil-men want to bring in factories and oil-leaks that stain the air and snow, and giant pipes that block the migration of herds of animals.

But they have had to deal with the Caribou People. For twenty thousand years the Gwich'in have been caretakers of this land, and for twenty years they have been fighting off the oil-men. Seven thousand of them live in a few small towns across Alaska and Canada. Sarah James is a member of the Gwich'in Committee. She says: 'We are the Caribou People. It's our clothing, our story, our song, our dance and our food. That's who we are. If you drill for oil here you are drilling right into the heart of our existence.'

Like the mythical flying reindeer that pull the sled across the skies at the winter solstice, the message of the Caribou People spread across America and Canada. They fired up millions of Americans and succeeded in pressuring the United States Congress into halting the proposed drilling in the Arctic. [3]

There is no need to drill this oil, to stain the snow and scorch the earth. There is a multitude of energising dances we can do with the sun, wind, water and bubbling heat from under the crust of our planet. I wrote some information sheets about this too. They are far sweeter to lie between than the fact sheets on the damage to the Earth.

I spoke to a few people who have invented technology that restores the gentle dance with fire and could stop the rapid downhill drive into a wall of extinctions.

I chatted to Professor van der Merwe, part of Professor Albert's research team at the University of Johannesburg. They have patented a solar photo-voltaic panel that is thinner than a human hair, and which can be produced at a fraction of the cost of previous panels.

I came across a buried story about a technology company, which claims to have developed a small affordable device that you can add to your car that converts water to gas, which then fuels the engine.[4] I phoned the inventor, Peter Jansen.[5]

'Is it true?' I asked.

'Yes,' he answered.

'You make me cry with happiness,' I said.

'We cannot keep living on the planet this way,' he stated, 'or it will soon be the end of us.'

He was not ready to talk much, as the product was not yet publicly launched. I felt concern for his life, as I have seen many examples of the ruthless ways oil companies have suppressed alternative technologies. He seemed well aware of this risk.

When I mentioned the billions of dollars the oil industry was making, he added: 'Billions *a day*.'

We agreed we would talk again, when the time was right. Hope and worry flutter in my heart and throat like a giant moth.

These fire dogs and inventors have shown me ways that we can dance with fire without burning the Earth. I am still finding my feet in this dance. I am doing a slow salsa with the fire inside me: feeling what lights up my heart and what heats up my feet; learning when to ignite and when to glow quietly; moving to the music of pain and joy, running at the enemy and playing with friends.

I want to dance! I have had enough of swinging tennis balls and tentative steps: it is time for my trial by fire.

It is a quiet evening and a group of us are standing on the beach, our feet on soft dry sand. The sun has set, but a breath of fire still paints the air. Blood red, gold and pink are smudged across the clouds ripped and stretched on the horizon, where the darkening sky meets the wide flat sea.

The Fire Dogs of Climate Change

Two dogs are lying quietly with their chins on the ground, watching us. In each hand we hold a chain that hangs down to a cloth oil wick. The others are old friends with fire, but tonight is the first time for me.

We form a large circle to initiate me into the dance. The man beside me ignites his wicks, bestowing on them his good wishes. Using his fire, he lights the chains of the woman on the other side of him, who adds burning blessings and passes them on. When the fire circles to me it is alive with the stories of all their dances. The tips of my chains bloom tangled petals of flame. I step into the centre of the burning circle and we spin our fires in unison. The tiny comets blaze past my feet and ears and there is a pulsing roar of the hot breath of our fires. From the cocoon of flame they swing armfuls of warmth towards me, as I move to the song of fire.

We are rivers and waves of energy, spinning and ducking and rising. We paint a flaming picture in the air of what is real and what is possible. For all the world to see.

One dog scratches its ear, then settles into the sand with its nose buried under its tail. The other sits straight up, points its nose at the burning stars and barks up into the sky. We dance with fire, and it dances us. The dance of destruction and creation. Stillness and movement. Fighting and building. Grieving and celebrating.

To tell you these stories, I ignited my pen. I danced with my heart and my ideas. I wove words and images; painted pictures in the air. For all the fire dogs to see.

Inspiring Examples: Groundbreaking Technology

There is sooooo much cool and brilliant renewable energy and energy efficient technology out there.

This is the section that should make the technical fire dogs prick up their ears, but the non-techies' eyes may begin to glaze over. I admit to being somewhat technically challenged myself, so I enlisted the help of techno-whizzes to research this section. Nevertheless, I find this stuff fascinating. I've tried to present it in a fun fashion, though I need to include some hard data for those who do know their stuff.

Many individuals and organizations have drawn up national or international programmes to replace fossil fuels with renewable energy (RE) and energy efficient (EE) technology [see, for example, Monbiot (2006); Greenpeace/EREC (2007) and Zero Carbon Britain]. [1]

I urge you to locate and look at the proposals most relevant to you. What I have documented here is not the mainstream RE and EE technology that these experts are advocating, but some of the *lesser-known* groundbreaking technology that is available now (or very soon), which could take these proposals *even further*. I have listed just a few inspiring examples – by no means a comprehensive catalogue.

A lot of this technology is as old as the hills (and as the earth, wind and sun), but has not been in mainstream production; other technology is brand new. I give examples of RE and EE technology rather than practices. However, energy-efficient practices can cut emissions dramatically, and using less energy is a vital first step on the path to sustainable energy use.

The costs of many RE technologies are dropping fast, while fossil fuels are spiralling upward. As Al Gore stated (in his speech calling for 100% of US electricity to come from RE in the next ten years):

'...the sharp cost reductions now beginning to take place in solar, wind, and geothermal power – coupled with the recent dramatic price increases for oil and coal – have radically changed the economics of energy.

'When I first went to Congress 32 years ago, I listened to experts testify that if oil ever got to $35 a barrel, then renewable sources of energy would become competitive. Well, today, the price of oil is over $135 per barrel. And sure enough, billions of dollars of new investment are flowing into the development of concentrated solar thermal, photovoltaics, windmills, geothermal plants, and a variety of ingenious new ways to improve our efficiency and conserve presently wasted energy' (Gore, 2008).

We need to override the fossil-fuel lobbies and quickly bring the best of the RE and EE technology (and practices) into mainstream use. Given the political will we could replace all nuclear, coal and oil throughout the world within the next ten years. And we could do it in a way that promotes the healthy development of all on this planet.

During wartime emergencies, nations have shown they can bypass the red tape, and dramatically and quickly shift production to make weapons of war. We are now under a threat that surpasses all world wars, and we need to produce the tools that can bring the Earth some peace.

Cars riding on water – and I've got one!

We've cut our petrol use by 24% overnight – with a little water! To be precise, it's not water alone that's done it – but a device that makes a hydrogen gas: HHO (known as Brown's gas or oxyhydrogen) from small quantities of water. There are claims that the use of HHO with petrol results in a dramatic reduction of all polluting emissions.

These HHO-generators produce 'hydrogen on demand', which is very different from running on tanks of hydrogen. The hydrogen for fuel that is being touted by some governments and businesses is fraught with problems: the production process is usually environmentally damaging, the storage is difficult and dangerous, and the costs are high.

If you remember even a little science (or spend any time in the kitchen), you know that a small amount of liquid makes a lot of gas. This HHO-booster uses one litre to make over 1800 litres of HHO. This is produced in increments, as required by the engine. The HHO is a fuel itself, but it also combines with the petrol, resulting in improved combustion and increased mileage.

Bowen fitted an HHO-booster into our Volkswagen Jetta, and after some fiddling, we are up and running. Every 500km, we top it up with 100ml of distilled water. We can't run purely on HHO because the process produces water vapour, which would rust the metal in our engine. Theoretically, a modified engine (perhaps made from ceramics?) could run on pure HHO.

Inspiring Examples: Groundbreaking Technology

In South Africa, this HHO device is made and sold mostly by backyard mechanics, but it seems to be sneaking into a more mainstream market. There are a number of varieties of this product, and it looks relatively simple to make. The Peswiki website lists a number of types and distributors, and also features the best DIY guide books on how to build your own HHO kit. Sale prices vary from about 100$(US) to 400$, but I reckon a clever woman could make her own for about 40$. HHO can be used in all cars: fuel-injected, diesel, hybrid, or carburettor-driven, but it seems like it works better in some than others. Reported results vary from 0 to 70% reduction of fuel use. One distributor told me that he was having much better results with the older cars than the new cars. The HHO-electrolyzer has a number of other potential energy-generating uses, such as making hydrogen for electricity-generation in fuel cells.

There are numerous low-emission vehicles coming on to the market (I describe some of these later on), and together with increased use of our feet, bicycles and public transport, these greener vehicles could really make a difference to transport emissions. But what is great about this little HHO generator, is that you can install it in your car and cut emissions *now*.

(www.peswiki.com; www.HHO4fuel.co.za)

My first experience of driving with the HHO-booster was strange and wonderful. The engine was quiet and powerful, and my sensation was of being lighter. I tried to work out why it felt so good. It was something to do with using water and air, and less oil. The Earth takes millions of years to produce heavy oil: so much time and algae and sunlight; it seems a sin to burn something that precious so quickly. But water and gas cycle freely in our atmosphere, and give so much energy with such little encouragement. Then I realized that the HHO-driving reminded me of something else that I had tried for the first time just the week before – on Muizenberg beach: the rush I had felt riding the air and water; another gift of renewable energy – *surfing*. Now my car was surfing too.

Fuel combustion agent could cut emissions by over 80%

I have just bought a bottle of a new product that could assist with a clean transition from fossil fuels to renewable energy. It is a combustion enhancer in liquid form that you add to your car fuel. Using nano-technology, this hydrocarbon solvent breaks the surface tension of the fuel, allowing a more complete burn. This dramatically improves efficiency, power and fuel consumption, and reduces engine wear. Most significantly, from a climate change perspective, distributors claim that it results in a reduction of harmful tail-pipe emissions (including CO_2) by at least 80%.

The Fire Dogs of Climate Change

The liquid is inexpensive (relative to the costs saved), at about US$3 per 50-litre fuel tank. Depending on your car, and driving conditions and habits, you should get fuel savings between 15-25% with petrol, and at least 15% with diesel.

This product is currently sold in South Africa as 'Naf-Tech', but future distribution may use other brand names.

I've heard and read testimonials from a few companies and people who have experienced good savings with Naf-Tech. I haven't had time to complete the tests on my own car, but I can tell you it didn't explode my fuel tank, which was a bit disappointing because Naf-Tech comes with a five million rand insurance policy (about 650 000 US$). It was tested by the South African Bureau of Standards (SABS) and is SANS 342:2006 compliant (i.e. acceptable for use in internal combustion engines). The South African trucking industry have tested and used it for over a year with great success.

If this technology really provides the reductions claimed, it should be a mandatory additive to all our petrol; allowing a glimmer of a smile to return to the face of the planet, without affecting the profit grin of the oil companies.

There are of course many other problems with the mining and processing of finite fossil fuels, and this product should be used to facilitate the clean transition to renewable energy forms, rather than delaying this transition.

(www.cleanliving.co.za; www.mxfuel.co.za)

Washing machines without water

Researchers at the University of Leeds, UK have developed a waterless washing machine! It uses less than two percent of the water and energy of a conventional washing machine. Plastic polymer chips are tumbled with the clothes to remove stains. The technology measures up to cleaning industry standards, removing virtually all types of everyday stains as effectively as existing processes, whilst leaving clothes as fresh as normal washing, and dry enough to dispense tumble-drying. It also means we could do away with toxic washing-powder and dry-cleaning agents.

That washing-powder jingle keeps going through my brain: *Good and clean and fresh, tra-la-la...*

(www.celsias.com) [2]

Inspiring Examples: Groundbreaking Technology

Fridges running on sound

Listen up, cool cats and hot dogs. Sound fridges, more accurately 'thermoacoustic chillers' are used by the US Navy, and were on display in an ice-cream shop on Earth Day in New York in 2004. Sound is used to compress helium – doing away with the chemical refrigerant and the compressor – making the fridges more energy-efficient and environmentally-friendly. Penn State, with support from Unilever were working on developing cheaper manufacturing methods for the 'green chillers.'

(www.thermoacousticscorp.com)

Heating and cooling using ground source heat pump

The ground source heat pump (GSHP) is a highly efficient way of cooling or warming buildings (and warming water geysers) and is used in many houses and office spaces in the USA, Scandinavia and other parts of Europe. Its heating efficiency is 50 to 70% higher than other heating systems and cooling efficiency 20 to 40% higher than available air conditioners.

A pipe circulates water from the GSHP unit to about 1.5 m under the ground – where the temperature is constant (e.g. 5°C). This condenses a refrigerant, which drives a stirling engine. This GSHP system generates between 2.5 and 4 times as much energy as it uses.

WhisperGen (a New Zealand Company) seem to be making the best GSHPs, using a newly developed and more efficient drive system ('Wobble Yoke') which replaces the traditional crankshaft system on stirling engines.

(www.gshp.org; www.whispergen.com)

LED lights the way

LED (Light Emitting Diode) light bulbs last 40 times longer than incandescent bulbs, and use about 13% of the electricity. They last ten times longer than the compact fluorescents and use about 60% of the electricity. LED has the added benefit of having no highly toxic mercury (unlike fluorescents). Single watt LEDs are cheap, but the light bulbs equivalent to the incandescent 60W (8 W LED) are currently very expensive (about $90). In the long run this is still cheaper on pocket and planet. We use soft LED 'low lights' in our hallway and toilet, which cost us about $9 (made by Osram). Prices could reduce dramatically with increased demand and government support.

(www.metaefficient.com; store.Earthled.com)

The Fire Dogs of Climate Change

Blown away by a kite ship

If you attach a giant kite to a heavy cargo ship, you can make fuel savings between 10-30%. Smaller ships can use conventional sails to cut fuel dramatically.

The heavy-lift project carrier 'Beluga Skysails' did a two-month test run with a 160 m² kite, from Germany to Venezuela, the United States and Norway. In moderate wind conditions, the SkySails-System pulled the ship with up to 5 tons of power at force 5 winds, amounting to 20% less energy use. This results in savings of about 2.5 tons of fuel and more than $1,000 a day. When the kite-sail is scaled up to 320 square metres, savings of up to 30% fuel are likely.

(www.belugagroup.com) [3]

Solar-powered boat crosses the Atlantic

On a sunny day in March 2007, the 'Sun 21' catamaran cruised into New York City. It had journeyed about 7000 miles across the Atlantic – using only solar power. This was the first solar-polar motorized crossing of the Atlantic. Michel Thonney, the skipper said: 'I don't particularly like speeches but I would like to shout out to the world – use solar energy!'

(www.transatlantic21.org)

Airships: planes powered by the sun

OK, these are not yet up and flying, but there are a number of companies working on it. The airship would replace the fuel-guzzling aeroplane. Similar in shape to a 'blimp', it has an outer shell filled with helium (to make it light). Its engines are powered by thin solar panels that drive motors that use super-conducting magnets in lieu of copper wire. There is also a backup engine. It has vertical take off and landing, reducing the amount of aircraft land necessary.

Keep a look out for identified flying objects from: Millennium, Aeros, Lockheed Martin, Techsphere Systems International and World Skycat.

(www.millenniumairship.com; www.treehugger.com) [4]

Internal combustion engines waste petrol

Engines may not make the world go round, but they make most things on it 'go.' And the one that most of us use to burn our fossil fuels is the 'internal combustion' engine. And guess what? It wastes most of the energy we put into it. After all those eons of the

76

Inspiring Examples: Groundbreaking Technology

Earth creating the oil, and the travail and damage to people and planet in digging and burning it, only 20-40% is translated into energy (most petrol engines are 20%). The rest is given off as heat or waste. So, let's look at some more efficient alternatives.

Electric motors are waaaay more efficient

Although electric cars may seem like modern developments, the electric motor was invented in 1888 (by Nicholas Tesla). Electric cars preceded petrol cars, but have repeatedly been squeezed off the market. Oil companies have squashed the electric car for obvious reasons, and motor companies have resisted them because most of automobile industry profit comes from replacing and repairing parts. The electric car has no bonnet full of oily mechanical parts (see www.whokilledtheelectriccar.com).

Electric engines are 80-90% efficient. So even if you are charging them from a fossil-fuel source, you are producing far less emissions. Of course, it would be best if you charge them from RE sources (if your electricity grid doesn't provide green energy, you could install your own).

Many motor companies are bringing out electric or hybrid vehicles (including Reva, Toyota, Honda, Chrysler, Volvo, KIA and others). I feature a couple of them below, but you do not have to rush out to buy a new electric car: you can get your existing car modified to a hybrid or pure electric vehicle. You may have to scout around, order an electric engine 'conversion kit' and find a qualified mechanic to do it, but it could be a lot cheaper (and produce less CO_2) than purchasing a new car.

In an experiment a couple of (*Mythbuster*) engineers gave themselves just 12 hours to convert a gas-guzzling racing go-cart to electrics. They succeeded and the electric one out-raced the fuel-based car (*Popular Mechanics South Africa (SA)*, June 2008).

As batteries improve, electric cars can travel longer distances without re-charging. You can charge them at home overnight, or do quick charges at plug-in points set up at garages. The batteries of electric cars can also be used to store energy from the grid to be used at peak hours in your home (which could take a load off energy demands).

Tesla electric sports car

The Tesla Roadster gets a special mention because it is so sexy (especially the red one!) and it shows off the efficiency and power of electric technology (which could be applied to cheaper cars). The Roadster is a fully electric sports car that can travel 220 miles (350 km) on a single charge of its lithium-ion battery pack and accelerate from 0-60 mph (97 km/h) in 3.9 seconds. The Roadster's efficiency is reported as 133 W·h/km (4.7 mi/kW·h), equivalent to 135 mpg (2.09 litres per100 km). Its motor efficiency is 90% on average

The Fire Dogs of Climate Change

and 80% at peak power. Tesla will also be bringing out a cheaper family electric car. Tesla Company is named after the genius inventor, Nicholas Tesla, who was known as 'the father of free energy.'

(www.teslamotors.com) [5]

Cute little Reva

This is the cutest little city car (check out the topless model), available in 2000 wild co-lours. At about US$10, 000, this Indian-made car is more affordable than most electric vehicles. Designed for city travel, it has a speed of 80km/h and a range of 80 km, on acid gel batteries. Reva are releasing another model shortly with lithium ion batteries that will have a greater range and speed.Reva is sponsoring the 'solar electric rock trip' across India (see 'Inspiring Examples: Action'), and this model will have a lithium bat-tery, with solar panels on the roof to top it up.

(www.revaindia.com)

Ultra efficient electric engine?

Zhaan Jordaan of South Africa has developed (and had independently tested) a proto-type of an electric motor that is 120% more efficient than conventional electric engines (*Engineering News, SA,* May 2008). Watch this space. The electric engine revolution is just beginning.

Cars that run on air

Compressed-air cars have been developed in France and India and are used as taxis in Mexico City and elsewhere. Like electric vehicles, they produce no tail-pipe emissions. Pistons are driven by compressed air rather than fuel. The energy required to compress the air is far less than the amount that would be burned by fuel.

The car can be filled with compressed air at a garage in a few minutes, or an air-compressor (in the car) can do the work while plugged in at home (this takes a few hours). As with electric cars, this could be done using a renewable energy source of electricity. Zero Pollution motors aims to bring out a 106mpg air-car in the US by 2010. The Air Car was developed by The MDI Group, which received funding from Indian car maker Tata Motors to build the car for the Indian market.

(www.theaircar.com; www.tatapeoplescar.com;
www.zeropollutionmotors.us)

78

Inspiring Examples: Groundbreaking Technology

Fuel-sipping cars

For exciting developments in car technology, have a look at the Progressive Automotive X competition. Ten million US dollars is the prize for the best 'fuel-sipping' mass-market vehicle. Innovative features of existing entries include: micro-generators (in the car) for recharging the batteries; re-generative braking; solar roof-panels and a fuel-vaporizing system.

(www.popularmechanics.com/drivegreen) [6]

Plan for 95% reduction of cradle-to-grave vehicle emissions

The carbon-emission reductions of a vehicle or a motor cannot be measured just by its use of fuel. The entire construction of the vehicle, its materials, maintenance, repairs and eventual destruction are all factors contributing emissions. An international vehicle design summit set up a 'global consortium to design, build and bring to market the VDS Vision 200, a hyper-efficient 4-6 passenger vehicle earmarked for India that will demonstrate a 95% reduction in embodied energy, materials and toxicity from cradle-to-grave.'

(www.vehicledesignsummit.org)

Electric bikes and scooters

Electric bikes are cheaper and even more fuel-efficient than cars. Of course bicycles are even better, but for those who don't have the leg power for those long distances or steep hills – there are electric bicycles, scooters and motorbikes. There are even fold-up electric bicycles – you can carry them onto public transport or into your office.

In 1996 the Chinese government banned petrol-driven scooters in major urban areas. Sales of electric bicycles in China are over a million a year – and growing (*Popular Mechanics SA,* 2003). The US electric bike prices I found were expensive, but should drop as demand goes up.

(www.metaefficient.com; www.greenspeed.us) [7]

Electra-flying!

An electric airplane flew at the largest air show on the planet, and wowed the crowds not with its sonic boom – but its silence. In August 2008, the small plane made three passes over the main runway at the EAA Air Venture Oshkosh.

Designer, Randall Fishman, received an award and a standing ovation from the crowd at an earlier ceremony. He stated that the quietness and the absence of vibration

The Fire Dogs of Climate Change

are two of the Electra-Flyer-C's best qualities. Not to mention the cost. You can recharge its 18hp motor in two hours with a 220-volt charger (or six hours with a 110-volt) at the cost of 75c (US). No, that was not a typing error.

The ElectricFlyer-C has a top speed of 90mph and a flight duration of up to two hours with a lithium battery pack. A small 110-volt charger can be carried as baggage and used to prolong the flight time. The Electric Aircraft Corp doesn't sell the plane, but does offer 'powerplant packages', including motor and controller. Jai Reddy, who lives not far from me in Cape Town, has developed an electric gyrocopter. He should have a prototype in the air in 2009. He has turned down offers to sell the patent, because he wants to make sure the invention gets off the ground and flies onto the market. He has a mechanical method of regeneration that will recharge the batteries by 30%, as well as solar panels to top them up.

There are other planes close to take-off. Advanced Technology Products of Massachusetts has designed a plane with a permanent magnet motor and batteries stored in the wings (*Popular Mechanics, SA*, March 2003). Sonex, as part of their 'E-flight initiative' should have something in the air by the time you pick up this book. Keep an eye on the sky.

(www.ElectraFlyer.Com; www.satoritech.co.za; www.sonexaircraft.com)

Electric engines for boats

Solomon technologies has developed a high-torque, low-rpm electric motor for boats from 20-55 feet.

(www.solomontechnologies.com)

Boat runs on steam

Here's an efficient, quiet, external combustion boat engine that runs on steam. The Pursuit Dynamics PDX marine drive works by injecting steam through a rear-facing, ring-shaped nozzle into a cylindrical chamber. As the steam emerges – at three times the speed of sound – it condenses rapidly, generating a shock wave that pulls water in through an intake and expels from the rear, thus generating thrust.

(www.pursuitdynamics.com)

Archimedes' screw principle drives boats and micro-hydro

The company, FLUID motive, Inc. has developed a replacement for a propeller that is named a 'Ribbon Drive propulsor'. It is based on the 'Archimedes screw' principle, and is much safer and more efficient than conventional propellers.

Inspiring Examples: Groundbreaking Technology

HydroCoil Power, Inc. is selling the Hydrocoil turbine, based on similar technology. Moving water spins the screw-shaped shaft ('vane') – which drives an alternator producing electricity. Large supplies of water are not required, due to its highly efficient design, which optimizes energy extraction.

This technology can be used in low-head hydro applications such as rooftop water towers, since it does not require high-pressured entry. Two or three may provide enough power for a typical home using 500-600 kWh/month, while 40 to 100 may be used together to provide the power needs of a small business.

(www.fluidmotive.com; www.hydrocoilpower.com) [8]

The massive yet tiny (MYT) mega-powerful engine

The Massive Yet Tiny Engine, or the MYT Engine has a phenomenal weight to power ratio. It is claimed to deliver 848 horsepower from an engine weighing just 150 pounds, which would give it a power-to-weight ratio 40 times greater than a standard internal combustion engine. The MYT Engine has the potential to replace all the existing internal combustion engines and jet engines (providing fantastic fuel-efficiency). A diesel working prototype was developed, but there is as yet no sign of MYT on the market.

(www.angellabsllc.com)

Stirling engine

The stirling engine is an external heat engine (as opposed to the conventional 'internal combustion' engine). It has a 'closed-cycle', containing a gaseous working fluid – permanently contained in the engine's system. It is quiet, efficient and can utilize renewable energy sources (e.g. sun, geothermal), or what would otherwise be waste heat.

Stirling engines are usually used in large industrial processes. However the FREEPOWER ORC Turbine Generator is a stirling-engine driven small generator that can be used in homes, cars and small industry to generate electricity. The system can be driven by any external source of heat, i.e. hot air, steam, hot oil, exhaust gases from engines, waste heat, solar thermal energy, and so on. The FREEPOWER system has been developed to run on low-grade waste heat. The generator comprises a small high-speed generator (similar to a car alternator), which is driven by a two-stage turbine built onto the generator shaft. The turbine is driven by high-pressure hot gas. Running on waste heat, the efficiency of the system – heat in to usable electricity out – is from 10% at 110°C to more than 22% at 270°C. This efficiency may seem low, but because it uses waste or renewable energy, the effectiveness is in fact high.

(www.stirlingenergy.com; www.freepower.co.uk) [9]

The Fire Dogs of Climate Change

'Green Revolution' Cyclone engine

Another great external combustion engine is the 'Green Revolution' Cyclone engine by Cyclone Power Technologies. It can be used for anything from a weed-eater to a big ship or a hospital generator. It can run on any fuel – liquid or gaseous. The external combustion chamber heats a separate working fluid (de-ionized water) – which expands to create mechanical energy by moving pistons or a turbine.

Initial tests of the engine used fuels derived from orange peels, palm oil, cottonseed oil, and chicken fat. Whereas almost anything can go into a Green Revolution Engine, almost nothing comes out. Lower combustion temperatures and pressures create far less exhaust gases.

(www.cyclonepower.com)

Waste to energy

Apart from the external combustion engines, there are a number of methods that convert waste to energy. I love this idea because it solves two (sometimes more) major environmental problems at once. And many of the technologies create useful 'byproducts' (e.g. compost or food) after the conversion process. Have a look at the Peswiki 'Waste to Energy' page. It presents a 'prioritized listing of the best resources and most promising up-coming technologies. Taking into mind such things as overall efficiency of converting waste to energy while neutralizing any toxicity that might reside in the waste.'

Waste to energy technology and methods can vary from small-scale home devices (e.g. biodigesters to provide cooking gas) to large-scale electricity production (e.g. from landfills or industrial heat and water waste). As with all technology the application can vary enormously in its ability to heal or destroy the environment. A good example of this is the HAWK 'waste to energy' microwave technology, listed amongst Peswiki's 'top 20' energy technologies.

(www.peswiki.com)

A huge supply of cheap oil from waste?

The Global Resource Corporation has produced a technology that uses high frequency microwaves to extract gas or oil from almost any carbon based waste. In a relatively cheap and simple process, they turn old tires, plastics and other carbon-based substances such as shale, coal, and tar sands into gas and oil. The energy input is way less than the potential energy output of the fuel, and the transformation process is free of CO and CO_2. The device can also be used in situ to extract fuel from sources previously

Inspiring Examples: Groundbreaking Technology

unviable (e.g. in abandoned coal mines and capped-off oil wells). It can be used to produce methane and hydrogen from low grade coal.

The process uses High-Frequency Attenuating Wave Kinetics or HAWK Microwave. It is patent pending, but the first commercial plant is under construction. The technology is being touted all over the world (including China). Global Resource Resource Corporation is a worldwide petroleum research, engineering, development, and manufacturing company offering solutions of enhanced energy recovery. Any alarm bells ringing yet? Can you see the oil and coal companies grinning?

This HAWK technology could be used to produce a huge supply of diesel type fuel, setting peak oil back by 100 years, and cutting the price of fuel. If it wasn't for the CO_2 problem, this could be great. We could get oil from waste and give the earth a break from all the damaging digging and drilling. But sadly there is no moratorium on extracting fossil fuels, and these fuels do produce high CO_2 emissions. If we are relying on market forces (and escalating fuel prices) to get clean green energy into production, this invention could be the nail in the coffin of renewable energy.

But, this same technology could be applied to producing gas and other low-emission fuels from waste matter. This will depend on the ability of our governments to regulate. Either way, I would watch this company like a hawk.

(www.peswiki.com; www.globalresourcecorp.com)

Biogas

Biogas digesters are a simple, well tested, but relatively underused technology that addresses the problem of sewerage disposal at the same time as energy production. The current disposal methods of human and animal waste are hazardous to the environment on many levels (for example there are large dead zones in the sea because of sewerage nutrient pollution). Biogas digesters can be used to capture the methane produced by livestock manure, which is a significant contributor to greenhouse gases.

Biogas can be made from a variety of waste (wet, dry, organic, fermentable). It can be used for cooking and heating, or in electricity-generating plants, to run fuel cells and engines (stirling or internal combustion) and in small-scale energy production.

Biogas digesters also produce useful residue. Depending on the waste that was processed, this can be used as a nutrient-rich fertilizer for food-gardens, animal-fodder, or fuel 'pellets' to replace fossil fuels.

Alongside other alternative heating and cooking sources (e.g. hot boxes and solar cookers), biogas reduces indoor smoke pollution, and protects local forests that are destroyed for fuel.

(www.renewableenergyworld.com; www.itpower.co.uk) [10]

The Fire Dogs of Climate Change

Sun converts waste to hydrogen

Solar Hydrogen Energy Corporation (SHEC) has developed a process that will convert landfill and other waste methane into clean hydrogen, using the power of the sun for the reformation, at a price comparable to traditional hydrogen production methods.

(www.peswiki.com; www.shec-labs.com/)

Hydrogen fuel from wastewater + bacteria?

Bruce Logan of Penn State University has developed microbial fuel cells that can turn almost any biodegradable organic material into hydrogen (with water as the only emission). Bacteria that feed on vinegar and wastewater are zapped with a shot of electricity, resulting in the production of hydrogen gas. It is a similar system to water electrolysis, producing oxygen and hydrogen, but this method is claimed to be more efficient. Because the bacteria do all the work, it uses about a tenth as much energy as electrolysis. You could locate these cell reactors in food and agricultural industries, where the waste can be used to make hydrogen.

Someone please make good friends with all these bacteria (put that antibacterial soap away!), and see what else they are willing to do. Perhaps – in alliance with the astounding algae – they can help us get it all under control?

(www.enn.com; www.keppelseghers.com) [11]

Fuel from algae (the CO_2-gobbler)

Algae double their mass several times in one day, using only sunlight and CO_2. It is CO_2-eating algae that altered the Earth's atmosphere (millions of years ago), creating the conditions for oxygen-sustained life on Earth today. Perhaps it has an equally significant role to play again. Today it spreads across vast areas of the sea – gobbling more CO_2 than all the rain forests. Ancient algae reserves are the source of current oil, and fresh algae could be source of biofuel if – and this is always a big if, when it comes to the smell of profits – it can be farmed sustainably and with environmental sensitivity. It is also extremely low maintenance, thriving in non-potable water on non-arable land and delivering 10 to 100 times more energy per acre than cropland biofuel.

Sapphire Energy, a San Diego based company, is at the forefront of the development of large-scale algae fuel production. Aquaflow, from New Zealand is trying to become the first company producing large amounts of fuel from wild algae instead of the genetically modified micro-organisms employed by some of their competitors.

Inspiring Examples: Groundbreaking Technology

If we need to use biofuels, this seems like a clever source, though I am unsure of the amount of CO_2 emissions produced by burning this fuel (and how this compares to the amount that it eats up in the growing process). I suspect it should be used in moderation. Nevertheless I would still like to nominate algae, as well as the germs mentioned above to run for presidency in the next US elections.

(www.celsias.com) [12]

Who are we to blow against the wind?

Wind power is currently the most used, and fastest growing, form of renewable energy. Even a nuclear scientist can work out how a windmill works. In the Northern hemisphere the wind energy price is competitive with fossil fuels. By the end of 2007, wind power was generating about 94 000 MW (megawatts) worldwide – three quarters of the RE electricity generated in 2007. In some countries and regions wind already contributes 40% of the electricity.

(www.wwindea.org) [13]

But are the current wind farms the only or best way to capture the wind? Have a look at these innovative ideas:

Think vertical and small

Vertical towers are quiet, have smaller visual profiles, are less damaging to birds and wildlife, and are easy to install. The vertical-axis wind tower from Windhaus Turbines of Ontario, Canada claims to crank out 50% more power than conventional, horizontal-mounted designs. A 30m model generates 5 000kW/hour in a 19km/h wind (*Popular Mechanics SA*, August 2004).

Jay Leno (US TV-show host) has set up a vertical wind turbine for his big workshop. In a 45km/hour wind it generates 10 kW.

Check out the vertical 'quiet revolution' turbines and the ring-shaped 'Swift Rooftop' designs in the UK: small, beautiful, quiet, and designed to capture turbulent wind – they are ideal for micro-generation in suburban areas.

(www.windausenergy.com; www.pacwind.net;
www.quietrevolution.co.uk) [14]

The Fire Dogs of Climate Change

Or think wide and flat?

There are also many marvellous renewable energy inventions that are not yet commercially available. For example, the (patent-pending) 'Wind Wing' invented by Gene Kelley. According to Kelley, the oscillating Wind Wing can produce much more energy, much cheaper than conventional turbines. How can we get RE inventions fast-tracked in the patent office?

(www.2energycorp.com/home) [15]

Let the sunshine in!

'Scientists have confirmed that enough solar energy falls on the surface of the Earth every 40 minutes to meet 100 percent of the entire world's energy needs for a full year.' – Al Gore, 2008

But how do we harness this energy? Many solar power technologies have been developed. As the costs of fossil fuel (to people, planet and pocket) rocket, solar technologies are getting stronger, cheaper and more efficient. Some solar technologies are competitive in pocket price to coal. There is, of course, no competition when it comes to the costs to people and planet – solar wins hands down.

Ultra-thin, efficient solar PV panels

Professor Vivian Alberts, of the University of Johannesburg, South Africa, has developed an ultra-thin solar photovoltaic panel, which is more efficient than conventional PV (photovoltaic) panels. It is less than one quarter of the thickness of a human hair, requiring a fraction of the materials and costs. A factory in Germany is making these panels, and there are plans to set up production in South Africa. These PV panels are great for home, (electric) car and commercial use (*Popular Mechanics SA*, October 2005).

There are a number of other thin film PV designs. See, for example, DayStar Technologies' Silicon-Free Solar Cells. Daystar's metal foil design is not vulnerable to silicon shortages. Striving to make solar energy affordable, production of this thin film design is being increased to 20 MW per year. AVA Solar will start production in 2009 on cadmium telluride thin film solar PV modules (developed at Colorado State University). Produced at less than US$1 per watt, the panels will reduce the cost of generating solar electricity to about the same as the conventional grid.

(www.johanna-solar.com; www.peswiki.com;
www.daystartech.com; www.avasolar.com)

Inspiring Examples: Groundbreaking Technology

Concentrated Solar Power (thermal)

To provide larger scale electricity, for energy-chomping industries, we need something more, well, 'concentrated'. Concentrated Solar Power (CSP) is a long-standing, tried and tested technology, which is generating renewed interest in recent years. In the US, Nevada Solar One (64MW) CSP came on line in 2007, and 250 MW solar 'towers' (Germany technology, piloted in Spain) are being planned in Namibia and Botswana. Mr Gaylard Kombani of the Botswana Ministry of Trade and Industry (MTI), said that the Power Tower was cheaper than conventional methods of generating electricity (MTI, July 2008: http://us.mti.gov.bw). [16]

In the US, Stirling Engine Systems (SES) has partnered with Southern California Edison to install a 500 MW plant that is expected to open in 2009. It will cover 4,500 acres of land with 20,000 large, dish-shaped mirrors – producing more electricity than all other present US solar projects combined. The solar dish (consisting of 82 mirrors) focuses the sun's rays onto a receiver that transmits the heat energy to a Stirling engine. In the future, SES will also be participating in the biogas and hydrogen markets.
(www.peswiki.com; www.stirlingenergy.com) [17]

There have been recent advances in Concentrated Solar Power design. For example, Fresnel technology, used in a CSP station in Spain. Fresnel's system makes reflecting curved surfaces from standard glass mirrors – making the raw materials very inexpensive.

So far, three methods of solar thermal power generation have been identified with the potential to generate electricity within the 10 kW to 1000 MW range. These are dish/engine technology, solar tower technology, and parabolic trough technology. In the latter, curved mirrors capture the sun's warmth, which is then used to heat thermal oil, pumped through an absorber pipe. This in turn is used to create steam to drive a steam turbine.
(www.renewableenergyworld.com) [18]

Gigawatts of Concentrated Solar Photovoltaic

There have also been note-worthy developments in the design and materials used in Concentrated Solar Photovoltaics (CPV). For example, SolFocus CPV systems claim to deliver the lowest Levelized Cost of Energy (LCOE) and the highest energy density of any solar-energy generating systems. By concentrating sunlight using innovative optics onto a small area of high-efficiency solar cell material, SolFocus systems dramatically reduce the amount of expensive solar material used in the system. The largest portion

The Fire Dogs of Climate Change

of the SolFocus systems is glass and aluminium, which are readily available and have proven field durability. The result is solar energy systems that are cost-efficient, reliable and scalable globally to gigawatts of energy production.

In another solar CPV invention, IBM scientists are using a large lens to concentrate the Sun's power, capturing a record 230 watts onto a centimeter square solar cell (CPV). That energy is then converted into 70 watts of usable electrical power, about five times the electrical power density generated by typical cells using CPV technology in solar farms. The IBM system cuts the number of photovoltaic cells and other components by a factor of ten. The trick lies in IBM's ability to cool the tiny solar cell. Concentrating the equivalent of 2000 suns on such a small area generates enough heat to melt stainless steel. But by borrowing innovations related to cooling computer chips, they were able to cool the solar cell from greater than 1600° C to just 85° C.

(www.solfocus.com; www.ibm.com/green) [19]

Nano what?

Then there is nano technology… Bowen is still trying to explain it to me. Its implications are hard to get the head around: it deals with very, very small stuff and has very big results. I shall leave you to do your own research, but here let me mention that it has potentially revolutionary implications for RE. For example:

At the University of California, San Diego electrical engineers have created experimental solar cells spiked with Indium phosphide (InP) nanowires that could lead to even more efficient thin-film solar cells of the future.

(www.renewableenergyworld.com) [20]

Australian and Chinese researchers at the University of Queensland's Australian Institute for Bioengineering and Nanotechnology made a breakthrough in growing Titania nano-crystals – material that could be used for cost-effective solar cells, hydrogen production from splitting water, and solar decontamination of pollutants. Professor Max Lu said that his team had now made such materials 'easy and cheap.'

(www.englishdaily.com.cn) [21]

Energy from the earth

Heat stored beneath the surface of the earth holds 50, 000 times the energy of all the oil and gas in the world. Geothermal energy is more consistent (less weather dependent) than most of the other RE sources. Geothermal plants in the US generate nearly 3000 MW of electricity, mostly from reservoirs that are at least 149° C (*Popular Mechanics SA*, March 2008).

Inspiring Examples: Groundbreaking Technology

Even 74° C water can drive a turbine!

In Chena Hot Springs, Alaska is an unusual 400KW power plant. It generates electricity from water at 74° Celsius – which has previously been considered to be too tepid for power production. It does this by using an air-conditioning refrigerant, which evaporates at that temperature and drives the turbine. There needs to be a temperature difference of 38° C between the source (hot water) and the sink (the refrigerant), for this system to work.

The power plant is saving the hot springs $1000 dollars a day of diesel costs (*Popular Mechanics SA*, March 2008). [22] The implications of the 74° C are exciting not just for the geothermal industry, but also for other electricity generating methods (e.g. solar power can easily heat water to this temperature).

Nuclear energy without radioactive waste?

The nuclear spin-doctors are always trying to sell off marginally-less-expensive and a-tiny-bit-less-dangerous technology – as if it is cheap and safe. But in truth, all the existing nuclear plants are *very* expensive, and in spite of assertions by the industry and by James Lovelock, there is no way around the radioactive waste problem. No matter how small the quantity, it is toxic for thousands (sometimes hundreds of thousands) of years.

So I read about this new clean, nuclear 'fusion' method (on the 'top 20' Peswiki list) with some scepticism. But I concluded that maybe, it's possible that, at last, the nuclear spin-doctors will have something that they don't have to lie about. And then we could all sleep more peacefully at night.

Nuclear *fission* (splitting apart) is what powers present day nuclear power stations (and makes nuclear bombs). Nuclear *fusion* (combining/fusing) is the same process that powers the sun and most other stars. But its not just the idea of imitating stars rather the bombs that makes for sweeter sleeping …

American physicist, Dr Robert Brussard, and his team at Energy/Matter Conversion Corporation, have spent nearly 20 years developing a revolutionary radiation-free 'Inertial-Electrodynamic Fusion Device.' It was developed under a US Department of Defence contract and has recently been made public.

Dr Brussard states that this device could mark an end to fossil and fission fuels. He says: 'This is the only nuclear-energy releasing process in the whole world that releases fusion energy and three helium atoms - and no neutrons. This reaction is completely radiation free.'

For those who understand the details: '… the fusion process takes boron-11 and

The Fire Dogs of Climate Change

fuses a proton to it, producing, in its excited state, a carbon-12 atom. This excited carbon-12 atom decays to beryllium-8 and helium-4. Beryllium-8 very quickly (in 10-13 s) decays into two more helium-4 atoms.'

Brussard's device received the 2006, Technology of the Year award.[23] Brussard claims his fusion methods 'can make power at about five-eighth of the cost of conventional plants, process steam at one third of the cost of others, make cheaper syfuels (including ethanol), destroy nuclear waste, make space engines commercial, etc.'

His Inertial Electrostatic Fusion offers: small, efficient power reactors, 1-3% the size of current magnetic confinement reactors; clean, radiation-free energy utilizing p B-11; relatively simple engineering with commercial viability in six to ten years; and relatively low cost (US$150-200 million from program inception to 100MW demonstration of clean power). The technology does not have weapons potential so government funding has dried up. Brussard is now looking for private funds to take the process forward.

I urge environmentalists who are technologically clever to check this out. Wouldn't it be nice if we could give the nuclear-addicts a big bone to chew on – with no hangover that poisons us all?

(www.peswiki.com)

What do we do when it's dark and still: the storage question?

The problem with wind and solar is that it's not always blowing and shining, whereas fossil fuels can be burned up any old time. This is especially handy at times of peak demand. So how do you store large quantities of renewable energy? Build a giant battery?

Batteries

The British arm of German utility E.ONAG is building a giant battery. The prototype will be the size of four large shipping containers and will contain the power of ten million standard AA batteries, capable of producing 1MW of electricity for four hours. [24]

I shudder at the thought of such a huge chemical battery, but there are existing and forthcoming inventions that are smaller, more efficient and more eco-friendly.

Lithium ion batteries provide about 120 watt-hours (per kilogram), compared to lead-acid gel batteries at 32 watt-hours. Have a look at the efficient lithium ion batteries produced by: A123; Altair Nano-technology (who claim to have solved the overheating problem) and Compact Power. The pending 'ultra-capacitors' from EEStor may produce 280 watt-hours per kilogram. This Texas company is working on an 'energy storage' ultra-capacitor device made from ceramics. It's not technically a battery because it doesn't use chemicals. It can allegedly charge within five minutes (from a substation) with enough energy to move a car 500 miles on about US$9 worth of electricity – about 45 cents a gallon. This is a really big deal for renewable-energy storage, and the electric motor industry.

Remember the nanos I mentioned earlier? US Massachusetts researchers (MIT) are developing a Nanotube Super Capacitor Battery: a battery based on capacitors that utilize nanotubes for high surface area, enabling near instantaneous charging and no degradation. They are estimating 5 years to commercialization.

Another new and efficient storage mechanism is disc-batteries: spinning discs in a vacuum on electro-magnetic bearings. The disc is a motor that stores energy (charge it up and it spins faster; it slows down as you remove the energy source). Beacon Power makes a variation of this technology – flywheels that are used to store electrical energy for 'smoothing out fluctuations.'

And then, of course, there are fuel cells, which are more efficient and lighter than lead-acid and cheaper than lithium ion batteries. But fuel cells are much more than just batteries and I give them special attention later on.

(www.energytechstocks.com; www.peswiki.com; news.cnet.com) [25]

Compressed air stores energy

Compressed air is an old method of storing energy. When electricity is available, it is used to compress air into an underground reservoir, then in peak times it is released to the surface and used to run motors. This method is used by an 11-year-old power plant in Alabama (US), and a 23-year-old plant in Germany, both in caverns created by salt deposits.

(www.eere.energy.gov) [26]

General Compression, a small firm based in Massachusetts, is taking another approach. Its windmill compresses air directly. This has the advantage of eliminating two wasteful steps: the conversion of the mechanical power of a windmill into electricity and its subsequent re-conversion into mechanical power in a compressor.

(www.economist.com). [27]

Pumping water

Another age-old storage method is to use spare energy to pump water to a higher reservoir, and let it run down when needed. Power is generated at an efficiency of up to 80% and can be stored for days. Worldwide close to 280 pumped hydro-storage instal-

The Fire Dogs of Climate Change

lations exist with a total power of about 90 gigawatts. But be careful when messing with water: large dams can destroy livelihoods and ecosystems.

(www.electricitystorage.org)

Hot salt

Then there is, of course (shall I give you a moment to guess?) molten salt. This has highly efficient thermal storage properties. Salt can be heated to 1050°F, which allows high energy steam to be generated at utility-standard temperatures (1650 psi minimum, 1025°F), achieving high thermodynamic cycle efficiencies. Salt can be used to store the heat from Concentrated Solar Power Stations.

(www.renewableenergyworld.com) [28]

What is a fuel cell?

Although fuel cells are cutting edge technology for the future, they were developed over 100 years ago. Fuel cells are clever devices that extract electricity from relatively simple chemical reactions. They have the capacity to store energy – like a battery – but also to generate electricity. The proton exchange membrane (PEM) fuel cell drives the technology; but there is ongoing research into other fuel cells (molten carbonate, solid oxide and regenerative fuel cells). Hydrogen has been extensively used in fuel cell research and development, but it is not the only type of fuel cell. Fuel cells could be used to power anything from a hearing aid to large power plants. Every major automaker and oil company has a fuel cell research effort under way (mostly hydrogen related) (*Popular Mechanics SA*, February 2003). [29]

Fuel cells can make electricity from zinc + air

The Alternative Energy Development Coorporation (AEDC) provided each home in Kwa Mpungose (KwaZulu-Natal, South Africa) with zinc-air fuel cells, providing 35 000 people with electricity. Fuel cell shops exchange and sell recharged fuel-cell anodes.

The electrochemical process in a zinc-air fuel cell is very similar to that of a PEM fuel cell (but refueling is different) and shares some characteristics with batteries. Oxygen converts to hydroxyl ions and water, and the hydroxyl ions travel through an electrolyte to reach the zinc anode. The process creates an electrical potential. By connecting a set of zinc-air fuel cells, the combined electrical potential creates a source of electric power.

AEDC have projects to develop an electrical bicycle, wheelchair, car and taxi – all powered by rechargeable zinc-air fuel batteries.

Inspiring Examples: Groundbreaking Technology

In the US, Powerzinc is working on quick-refuel zinc-air fuel cells for vehicles. Conventional recharging (as with most electric car batteries) takes hours, but with the Quick-refuel system, the discharge fuel cell can be replaced with a fresh one in minutes at a refuel outlet. They are also working on zinc-air fuel cell systems in which the fuel is automatically regenerated (*Popular Mechanics SA*, February 2003). [30]

Using fuel cells to generate clean green power plants

The company Cenergie has developed the technology and strategies to set up cost-effective clean, green municipal power plants. They are initiating the installation of their hydrogen based Alkaline Fuel Cell (AFC) systems in the UK, France, South Korea, Washington state and California.

Leslie Berliant, writing for Celsias describes the advantages of the Cenergie plan. I quote and paraphrase her article at length because it utilizes many of the inspiring examples of clean technology that I have described in this chapter.

'Dominant non-renewable energy production, such as turbines, cost about $400 per kilowatt hour to manufacture. Cenergie's fuel cell technology is priced at about $440 per kilowatt-hour to manufacture, and will decrease in price based on economies of scale. They estimate that in about 5-7 years, the price will be $250 - $300 per kilowatt hour to manufacture. In addition, their technology contains no noble metals (such as platinum, currently one of the biggest cost hurdles for all other types of fuel cells), is manufactured using recyclable commodity materials, is 80% refurbishable into more fuel cells, and they assert that the only emission is half a litre of pure water per kilowatt hour (in the form of water vapor). Their system would allow for independent localized energy sources, not reliant on grids or power lines.'

Cenergie fuel cells generate electricity from hydrogen (H_2) and oxygen (O_2). Water (H_2O) is released as hot vapour, which can be condensed into its liquid form or used as steam in a Combined Heat & Power (CHP) system.

'Because you need pure hydrogen for the fuel cells to work, Cenergie is sourcing clean hydrogen. Using zero-emission, non-incineration pyrolysis and anaerobic digester partner technologies they can convert locally collected municipal, commercial, industrial and agricultural waste into H_2 and commodity carbon ash, as well as using partner solar and wind technology to split water into H_2 and O_2 through electrolysis. Additionally, because a fuel cell also needs pure oxygen, the O_2 injection loop scrubs atmospheric O_2 clean of CO_2 and other particles. Rather than injecting the CO_2 into the ground for future generations to deal with, this captured CO_2 can be used to feed algae for bio-diesel, feedstock and fertilizer or in hot-house food production. A company called Novomer is currently commercializing new ways of utilizing CO_2 to produce high-

The Fire Dogs of Climate Change

performance, green plastics and polymers and is also examining the re-use of sequestered waste CO_2.

'Cenergie is also looking at steam reformation of landfill methane into hydrogen, essentially utilizing a resource that is currently being wasted and contributing to climate change. This waste gasification provides the lowest cost energy, coming in at $US .07 per kilowatt as opposed to .22 for direct hydrogen from fossil fuels (transported) and .15 for natural gas reformation to hydrogen. Methane reformation to hydrogen comes in around .08 to .09 per kWh and electrolysis pricings are currently being explored.

'Cenergie's strategy in the energy market is to keep the technology and sell the energy that they produce to cities and utilities in fixed price ten-year power purchase agreements which allows them to control maintenance, repair, replacement and material recycling. The long term price agreements are possible because the company is 100% vertical, uses commodity materials (making them less vulnerable to supply chain price fluctuations), and regularly recycles those materials into newer, more efficient fuel cells while keeping the materials from the landfill. The fuel stacks are supposed to be easily replaceable, quick start and modularly designed to avoid outages should any single stack fail. Most importantly, Cenergie asserts that they can be mass manufactured and price point competitive to current energy production.

'Right now, Cenergie is looking to raise a few hundred thousand US dollars to set up further operations in California where they already have a 90% complete Initial Manufacturing Plant in Riverside County and plan to have three Mass Manufacturing Plants throughout California that will produce over 750 megawatts per year per shift at the end of their 5-year ramp-up. This seems a small investment if indeed they are able to accomplish all that they promise: 24-hour constant base-load energy that is cheap, clean and silent, highly efficient, easy to operate, low temperature, flexible, recyclable and reliable, with zero emissions and no noble metals. Let's hope it's not too far away' (Berliant, January 2008, Celcias).[31]

Let's do more than hope – let's make this happen. [31]

(www.celsias.com; www.cenergie.com)

Ongoing RE and EE adaptations and advances

There are ongoing developments and modifications of products, processes and technologies that advance the efficiency of existing EE and RE technology (OK, non-techies, now you can let your eyes glaze over).

Some of these include:

- Superconducting (ultra-thin, graphite) wire – for efficient long-distance transfer of

Inspiring Examples: Groundbreaking Technology

electricity (see Monbiot, 2006, p104 and *Popular Mechanics SA*, November 2006). This could also be used to make more powerful electric motors;

- Improved insulation, e.g. with Aerogel, the 'lightest solid known' (*Popular Mechanics SA*, July 2003);
- A solar boiler with aluminium tubing (www.redrok.com/engine.htm);
- A scroll generator adapted to be an air compressor. It uses few moving parts, is efficient and produces (localized) heat that can be used by heat energy systems (i.e. it can become a generator) (www.copelandscroll.com);
- Magnets in motors that can maintain performance up to 200° Celsius, which have been developed by the US Dept of Energy, Ames Lab in Iowa (most magnets in electric motors lose half their power at 120° degrees) (*Popular Mechanics SA*, May 2008);
- Using lattice rather than solid tungsten filaments, which results in significantly greater efficiency in transferring heat to electricity (www.innovations-report.com);
- Efficient cost-effective construction of complex 3D solids, with 'fast-track fabrication' (*Popular Mechanics SA*, April 2008); [32]
- Developing alternatives to concrete (which has high CO_2-emissions). E.g. in China they are using bamboo for bridge construction. The bridges, built by a team of eight workers, can carry at least eight tons and last 20 years. Bamboo is fast-growing and eats up CO_2 (*Popular Mechanics SA*, May 2008);
- Simple (kelp-like) designs for wave technology that should result in lower costs, maintenance, and visual and environmental impact than other wave technology (www.bulgewave.com). [33]

Technology exhibition at UN Climate Change Conference

At the United Nations Climate Change Conference (COP 14), in Poland, December 2008, the Polish Ministry of Environment is hosting an exhibition of practical innovations, technologies and installations for climate change mitigation. Have a look at their online exhibition site.

(www.cop14.gov.pl)

The best website for groundbreaking technology

For ongoing developments and advances in RE and EE technology, the best resource I found was www.peswiki.com. It is a publicly editable site about Pure Energy Systems (PES), and highlights new developments in clean, practical, renewable energy solutions. Peswiki is guided by the New Energy Congress, a network of more than forty

The Fire Dogs of Climate Change

energy professionals who are dedicated to clean energy technology advancement. The site contains hundreds of topics, thousands of features, and wide coverage of energy inventions, companies and inventors. It claims to be the 'best directory worldwide on extreme cutting edge technologies'.

The site contains a list of the top 100 leading new energy technologies, selected by the New Energy Congress, according to these criteria: 'renewable, environmentally safe, affordable, feasible, credible, reliable, developed, safe and not encumbered by the politics of science'. I have covered a number of the technologies that they listed amongst their 'top 20.'

(www.peswiki.com)

Free energy website

As luck would have it, I only found this 'free energy' site after months of research, and about two minutes before the publisher's deadline. This is just as well, or else 95% of this book may have been dedicated to technology.

In his 'practical guide to free-energy devices', Patrick Kelly introduces you to a number of devices that tap 'zero-point energy'. These are techniques and technologies that allow you to get out more energy than what you put in. For those of you who really do want to find technological solutions (and are prepared to override your 'that's impossible' reaction), I urge you to set aside an hour or twenty to check out this website.

(www.free-energy-info.co.uk)

Suppressed technology: 200 mpg?

Most inventions do not make it to market. Some (not all) of these are because they are brilliant concepts that have been suppressed by those threatened by their production. The oil companies have bought up barrels of EE and RE patents, which sit gathering dust on their shelves. It's hard to get reliable information on the amount of suppressed technologies because it's been, um, suppressed.

A recent case of suppressed technology is recorded in the movie, *Who killed the electric car*? It shows how General Motors sabotaged their own sales and literally crushed their own cars.

The repression of the vaporizing carburettor is another story that is fairly well documented in auto-history. There is evidence that Charles Pogue (Canada, in the 1930s) and Tom Ogle (US, in the 1970s) both invented carburettors that could reduce fuel consumption to approximately 200 mpg! (This is equivalent to 85km/l or 1.2 litres per 100km.)

Inspiring Examples: Groundbreaking Technology

Pogue's invention was demonstrated by the Ford Motor Company and was headlines across the US. The claims were verified by members of the public who bought the carburettor. There was a dramatic dip in the oil share stock market. Pogue's carburettor ran on pure 'white' petrol without additives. The oil companies introduced additives to the fuel (that are still used today), which made the carb ineffective. The carb was taken off the market, and Pogue refused to discuss the matter with journalists. He allegedly acquired a large sum of money and moved to Montreal where he opened a company manufacturing oil filters.

Tom Ogle from El Paso, Texas, exceeded 200mpg with his 'petrol injection' device. He was allegedly offered huge amounts of money to keep it off the market, but decided to go ahead. He was shot and wounded shortly thereafter, and then died of an overdose of Darvon and alcohol a few months later. No suicide note was found.

There are hundreds of vaporizing carb patents currently owned by oil companies.

Those who would like to build Pogue and Ogle's inventions: for about one US dollar plus postage, you can get the details from the US patent office. Write to: The Commissioner of Patents and Trademarks, Washington D.C 20231, USA.

The patent number for the Pogue Carburettor is: 2026798 (January, 1936). For the 'Oglemobile', it is: 4177779 (December, 1979). These patents would have expired by now, so they are in the 'public domain' for all to use – the same may apply to a lot of the other suppressed patents (Auto Engineering and Spares, July/August 2003). [34]

One of the contestants in the Progressive Automotive X fuel-sipping competition (Fuelvapor Alé) is using a 'fuel-vaporizing system' and claims to get 2,55 litres per 100km. [35] Perhaps this technology is making a come back after all...

What will they think of next?

Lots of things! We just need the resources and laws that will allow the research and production to happen. Pronto! I have just given you a taste of what's available. I do encourage the techno-boffs to get to your workbenches and join those Watts to those Whatevers; and the activists to get to the streets and the meets. Together, we could break free of the profit-snares, and ensure the planet-saving technology is brought into mainstream production.

I think we may see inventions in the next ten years that will make teenagers gasp, and our grandmothers faint...

Inspiring Examples: Sustainable Living

Political organizations, actions and technology are all crucial in ensuring climate change mitigation. But organizations and technology will not take us far, unless we know where we are going. We need to be clear not only about what we want to destroy (e.g. fossil-fuel dependence) but also about what we want to create. There are people who have already begun developing and practicing sustainable ways of living. They are creating real examples of what is possible. They are building tomorrow, today.

On an economic level, I believe it is the practice of gender-sensitive and environmentally-conscious democratic socialism that can best meet the needs of the Earth and the people on it. Unfortunately, I do not think that we will be able to achieve this goal in time to save the Earth, so we also need to see what is possible within the current system. We need to find ways of reducing the fossil-chomping nature of the wealthy; and we need to look at ways of improving the quality of life of the poor majority on this planet – without spewing out more greenhouse gases.

In this section I present specific sustainability projects that set examples for us. However, there is also a lot to learn from the practices of the millions of working class and rural poor whose destruction to the Earth is very small relative to their numbers. It is also worth studying indigenous hunter-gatherer societies, who developed tools and practices (social, technological and spiritual) to live productively, in harmony with the Earth. Some of this knowledge is still alive and practiced today.

Many of the governments of the developing countries argue that they need fossil fuels to 'develop'. Some of the examples below illustrate that renewable energy can be a far better option for development – for both the poor and the planet.

It is up to governments to regulate and enforce emission reductions, but these examples of sustainable living practice give us some idea of how life *can* be lived in a friendlier way to the Earth.

Inspiring Examples: Sustainable Living

Here are some of the stories of the fire dogs that are not just barking, but doing...

Findhorn ecovillage

The Findhorn Community is one of the oldest, wisest and freshest examples of an eco-village. At Findhorn, people address sustainability not only as an environmental issue, but also in social, economic and spiritual terms. Their values are manifest in their beautiful buildings and gardens, wastewater treatment, organic food production, consensus decision-making, wind turbines and solar PV panels.

As well as sustaining a 500-strong resident community, Findhorn is a humming hub of international spiritual and environmental conferences, networking and education projects.

According to a 2006 study, Findhorn has the lowest ecological footprint for any settlement ever measured in the industrialized world – at about 50% of the UK national average. In specific areas it is even lower: the 'home and energy' footprint is 21% (feed-in renewable energy systems – it sells electricity to the grid); the food footprint is 37% (largely home-grown, organic, vegetarian and seasonal diet); and car mileage is 6% of the national average (car-pooling and high employment level within the community).

To check out some of the marvellous happenings at Findhorn, go to Findhorn's website and and to Jonathan Dawson's weekly blog. [1]

Findhorn is one of many eco-villages across the world. Have a look at the Global Ecovillage Network to read more about the numerous 'centres of innovation and inspiration, introducing new technologies and social systems that spread out into the wider society.'

(www.findhorn.org; www.gen.ecovillage.org)

Transition Towns network

'You never change things by fighting the existing reality. To change something, build a new model that makes the existing model obsolete.' – Buckminster Fuller, cited on Transition Town Totnes website

The Transition Towns network provides a model of change for towns responding to the challenges of peak oil and climate change. They suggest mechanisms by which a community can work together to 'unleash the collective genius of their own people' to drastically reduce their carbon emissions. The founder of the UK-based movement, Rob Hopkins, outlines their approach in *The Transition Handbook: From oil dependency to local resilience*. The subject of their website is serious, but their style is lots of fun.

The Fire Dogs of Climate Change

Already there are over 60 communities around the world that have been inspired to become an official Transition Town, City, Village or area, with 700 others mulling it over. Community representatives in my own coastal suburb of Muizenberg, Cape Town are amongst the 'mullers'.

Totnes became the first Transition Town in 2006. Totnes have a number of groups looking at everything from 'buildings', 'energy' and 'local government', to 'education' and 'heart and soul'. Their 'Energy Descent Action Plan' involves finding ways to reduce the current nine barrels of oil per person per annum (current UK average) down to one barrel (or less) per person by 2030.

They are implementing a range of projects including: their own local currency, composting toilets, low energy street-lighting, promoting local produce, effective garden use, renewable-energy electricity, nut tree planting, cycling paths, and 'story-telling the future to educate and inspire'.

'Transition Town Totnes believes that only by involving all of us – residents, businesses, public bodies, community organizations and schools – will we come up with the most innovative, effective and practical ideas, and have the energy and skills to carry them out. Our future has the potential to be more rewarding, abundant and enjoyable than today, and by working together we can unleash the collective enthusiasm and genius of our community (that means you!) to make this transition.'

(www.transitiontowns.org; www.totnes.transitionnetwork.org)

Urban carbon management

In addition to Transition Towns, there are numerous climate change initiatives in urban areas around the world. For case studies of strategies and programmes from Mexico City to London to Shanghai have a look at the urban and regional carbon management website.

(www.gcp-urcm.org) [2]

Feed-in tariffs give you a check instead of a bill

A strategy that has been implemented in many European countries is 'feed-in tariffs', that allow households and RE companies to sell their renewable energy back into the central electricity grid. At the end of the month, households and companies receive electricity checks rather than bills. Tariffs can be used to subsidise and encourage renewable energy use and production.

One of the countries to recently implement this practice is Switzerland. In 2008, Swiss federal government launched a full system of feed-in tariffs differentiated by

100

Inspiring Examples: Sustainable Living

technology, size, and application. There are tariffs, or payments per kilowatt-hour (kWh), for solar photovoltaics, wind, hydro, geothermal, and biomass. The Swiss system, like those in Germany, France, and Spain, pays a renewable energy generator for every kWh of electricity generated.

(www.wind-works.org)

Institutions sharing ideas and training

There are a number of inspiring individuals, institutions and networks that are sharing ideas and practices about how to live well, and in accord with the Earth. Some of them are educational institutions. I list a few of them below:

Gaia Education and the GEESE

Gaia Education develops courses on sustainable community design and development, The team of ecovillage-based educators are known as the GEESE: Global Ecovillage Educators for a Sustainable Earth. They draw on the experience and expertise of some of the most successful ecovillages and community projects across the Earth.

(www.gaiaeducation.org)

Wiser Earth

Wiser Earth is a community directory and networking forum that maps and connects NGOs and individuals addressing the central issues of our day: climate change, poverty, the environment, peace, water, hunger, social justice, conservation, human rights and more. Their website features over 100, 000 organizations, groups and individuals involved in aspects of sustainable living.

(www.wiserearth.org)

Bioneers

Bioneers is a forum for connecting the environment, health, social justice and spirit with a broad progressive framework. They are committed to finding practical solutions for people and planet.

(www.bioneers.org)

101

The Fire Dogs of Climate Change

CIFAL Findhorn

CIFAL Findhorn – the only UN-affiliated training centre in Northern Europe – is based at the Findhorn Ecovillage. It operates as a hub for training, capacity-building and knowledge sharing between local and regional authorities, international organizations, the private sector and civil society on all aspects of integrated sustainable development, and other global goals of the United Nations.

    (www.cifalfindhorn.org)

Ocean Arks

The mission of ocean arks is to disseminate the ideas and practices of ecological sustainability throughout the world. Their motto is 'To Restore the Lands, Protect the Seas and Inform the Earth's Stewards'. See their list of new publications on ecological design and Dr. John Todd's Comprehensive Design for a Carbon Neutral World.

    (www.oceanarks.org)

Worldchanging

Worldchanging was founded on the idea that 'real solutions already exist for building the future we want. It's just a matter of grabbing hold and getting moving.'

    (www.worldchanging.com)

Centre for Alternative Technology

CAT aims to offer practical solutions to 'some of the most serious challenges facing our planet and the human race, such as climate change, pollution and the waste of precious resources.' The key areas they work in are renewable energy, environmental building, energy efficiency, organic growing and alternative sewerage systems. They aim to show that 'living more sustainably is not only easy to attain but can provide a better quality of life.'

    (www.cat.org.uk)

Schumacher College: transformative learning for sustainable living

Many of the inspiring projects around the world have the participation of people who have trained at the Schumacher College, in the UK. 'The College is renowned for the excellent teachers that lead its courses. People come to Schumacher College, in the heart

Inspiring Examples: Sustainable Living

of the Devon countryside, to discuss sustainability. What and how they learn stays with them for a lifetime.'

(www.schumachercollege.org)

New Economics Foundation: living well need not cost the Earth

New Economics Foundation (NEF), an 'independent think-and-do tank', was founded in 1986 by the leaders of The Other Economic Summit (TOES), which forced issues such as international debt onto the agenda of the G7 and G8 summits. They are creating an economics in which 'people and planet matter.'

They aim to improve quality of life by promoting innovative solutions that challenge mainstream thinking on economic, environment and social issues. NEF combines rigorous analysis and policy debate with practical solutions on the ground, often run and designed with the help of local people. They believe that 'living well need not cost the Earth.'

(www.neweconomics.org)

Ideas worth spreading

'TED, ideas worth spreading' collects inspired talks by the world's greatest thinkers and doers.

(www.ted.com)

Diet for a small planet

The books and website of Frances Moore Lappe, US author of Diet for a Small Planet have inspired many people.

(www.smallplanet.org)

Hope Building

Hopebuilding Wiki was created 'to share stories of achievement by ordinary people who are doing extraordinary things to make their world a better place to live in, but whose stories are not as widely known as they should be.'

(www.hopebuilding.pbwiki.com)

The Fire Dogs of Climate Change

Award-winning climate change projects

Ashden awards celebrates and rewards 'visionary champions who are finding solutions to climate change that are also bringing real social and economic benefits to their local communities. Across the UK and developing world, our award winners are inspirational examples of simple, practical ways to cut CO_2 emissions while also improving quality of life. Whether harnessing technology, energy efficiency or renewable sources such as solar, wind or biomass they're all beacons that we use to encourage others to take the sustainable energy path.'

Examples of these include a 'Fruits of the Nile' solar fruit-drying project in Uganda, a Technology Informatics Design Endeavor (TIDE) project making wood-saving stoves in South India, and a community wind project in Scotland.

(www.ashdenawards.org) [3]

Poor to sell (biogas) electricity to the rich

Energy Forum is using biogas to generate cheap off-grid electricity for villages in the Dry Zone in Sri Lanka. They hope to model this technology for wider replication.

In Bangladesh, the University of New South Wales is using a new finance model for RE technologies. 'An implementation agency will assist rural poor villagers in the business to sell electricity to wealthier members of the village. Poor people will get ownership of the technology after the payback period of the technology. In this project, small biogas plants connected to latrines will produce methane to generate the electricity for the rural costumers.'

The poor selling to the rich? This makes a refreshing change from the usual patterns of the fossil fuel industry.

(www.energyforum.slt.lk; www.unsw.edu.au;

www.wisions.net) [4]

Water wheels in the Amazon to provide electricity

An organization in the Amazon is installing Low Head Micro Hydropower in two villages in the Tapajos region. 'The principle of this innovative technology is to apply broad water wheels with a small diameter to the low water levels of creeks on the river.'

This could provide electricity 24 hours a day, and will replace the expensive and polluting diesel generators that currently provide energy three hours a day.

(www.wisions.net)

104

Inspiring Examples: Sustainable Living

Handbook on participatory development of micro-hydropower

Many others are using micro-hydropower to provide hydro-electricity. This is much friendlier to people and the environment than large dams. For example ADEID (Action pour un Dévelopement Équitable, Intégré et Durable) has had 15 years of experience in the participatory development of micro hydropower plants in rural areas of Cameroon and other African countries. They are producing a handbook to share the lessons they have learned.

(www.adeid.org; www.wisions.net)

Non-profit coop to run wind energy project on Pacific Island

In Vanuatu, an island country in the South Pacific Vanuatu Renewable Energy and Power Association (VANREPA) has initiated a range of renewable energy projects, including wind power, solar power, solar desalination and micro-hydro technologies.

Vanrepa has launched its project 'The Answer is Blowing in the Wind,' on the islands, Aneityum and Futuna. It will begin with providing electricity for schools and other institutions, with a longer-term goal of 100% renewable energy on these islands. Vanuatu is classified by the UN as a 'least developed' country, and most of its 200,000 population live in remote rural areas, and are engaged in subsistence agriculture.

VANREPA aims to establish a Renewable Energy Service Cooperative that will provide the necessary technical and management support. The coop as a non-profit will sell renewable energy to end-users. This organization is seen as essential to ensure the sustainability of the project, as it is 'by strength of its management and support, rather than by strength of its technology' that a project succeeds.

(www.vanrepa.org; www.wisions.net)

Practical action to reduce poverty

Practical Action (an initiative of The Schumacher Centre for Technology & Development) uses sustainable technology to reduce poverty in developing countries. They are currently implementing over 100 projects worldwide. Combined with their consultancy and educational work they outreached to about 664,000 people in 2006/2007. In 2008 they won a UNEP prize for a project in the Eastern Andes, Peru, in which they set up 47 micro-hydro schemes bringing clean power to about 30,000 people.

(www.practicalaction.org)

The Fire Dogs of Climate Change

Barefoot rural woman make great solar engineers

The Barefoot College in Rajasthan, India trains rural women and youth in a range of practical technical and ecological skills. One of their projects is the training of women as 'barefoot solar engineers' to fabricate, install, maintain and repair solar PV systems.

So far, they have solar-electrified at least 300 adult education centres, 870 schools and 350 villages (12,000 households). Their own college (which spreads over 80,000 square feet) was built by barefoot architects and is completely solar-electrified.

The Barefoot approach to solar electrification 'identifies indigenous knowledge and vastly under-utilized practical wisdom of the poor, upgrades their basic skills, builds up their confidence (they already have the capacity), and applies it for their own development. It builds the confidence of villagers from the very beginning, in a non-hierarchical learning environment based on learning-by-doing.'

Women are preferred to men because they are generally more stable, more likely to stay in the villages, and they usually teach other women what they have learned.

Barefoot College trains women not only from India, but also from Africa. These women have shown that language, illiteracy, and culture are no barriers to practical mastery of solar PV systems. 'As the College ramps up its solar electrification projects across India and Africa, these women are leading by example – showing how the skills of the rural poor can drive their own development.'

(www.barefootcollege.org) [5]

World Bank projects

World Bank and International Monetary Fund policies have wreaked social, environmental and economic destruction for decades, so I found myself balking at looking at the World Bank's list of 'inspiring and replicable' examples of sustainable energy projects. The World Bank is notoriously full of contradictions. Nevertheless, they may at times fund people and projects that get up to good things, so go to their website, and see for yourself. Have a look at the book (mentioned on this site) by Paul Osborn: *Sustainable Energy: Less Poverty, More Profits.*

One promising project, which the World Bank initiated (together with US Aid, the US Department of Energy, the National Renewable Energy Laboratory, Winrock and other private companies) is the Global Village Energy Partnership, which is currently hosted by Practical Action (UK). GVEP helps developing countries to set up energy action plans.

(www.worldbank.org/astae; www.gvepinternational.org)

106

Inspiring Examples: Sustainable Living

BEN and the bicycle

'Adding highway lanes to deal with traffic congestion is like loosening your belt to deal with obesity.' – Louis Mumford, city planner, cited in BEN report, 2007

Bicycles are healthy for your legs, heart, pocket and planet. Inexpensive and easy to repair, you can ride them to work and ride them to play. There are many organizations around the world that promote the use of bicycles. One of them is the Bicycle Empowerment Network (BEN) in South Africa. The mission of BEN is poverty alleviation through the use of bicycles. Together with local and international partners, BEN gets (often second-hand) bicycles from Europe, the Americas and Asia to Southern Africa; establishes community-based bicycle workshops; and sets up bicycle paths.

(www.benbikes.org.za) [6]

No till farming reduces CO_2 emissions

Agriculture is a significant contributor to CO_2 emissions. One of the reasons for this is the CO_2 released into the air by land disturbance.

Repeated tillage also destroys the soil resource base, causing adverse environmental impacts. Tillage degrades the fertility of soils, causes air and water pollution, intensifies drought stress, destroys wildlife habitat, wastes fuel energy, and contributes to global warming. The no till (or zero till) 'conservation method' is increasingly being used by big farmers because it improves soil quality, producing better crops.

South African Dirk Lesch, Swartland Canola Farmer of the year, produced more than double the yield of the average farmer in the same district. '*Ja*,' he says, 'the plough died in 1989, when Pierre Matthee… beat my yield on my own land with a no-tillage experiment block of wheat.'

The carbon content in Dirk Lesch's land is more than three times the level in neighbouring farms that practice ploughing (Farmers Weekly, June 2008).

In a properly designed no-till system, pest (weeds, disease, and insect) control is accomplished primarily with the following cultural practices: rotation, sanitation, and competition.

(www.no-till.com)

Organic, free range and permaculture farming

There are many other agricultural techniques, technologies and methods that are productive, friendly to the Earth and reduce CO_2 emissions. Many of these have been developed in the practice of free-range, organic, biodynamic and permaculture farming. These practices work together with nature, and do not use (petroleum-based) fertilizers and toxic herbicides and pesticides. They minimize waste, GHG emissions and energy use. They also challenge, and provide alternatives to, the existing methods of (feed-lot) meat farming. Earth-friendly technologies have been used for centuries by farmers around the world, and have been developed and documented in modern literature.

For a great overview of agricultural practices that can reduce GHG emissions, see the chapter 'You the Farmer', by Linda Scott and Leonie Joubert, in *Bending the Curve*, Zipplies (ed.) (2008).

In her book *Animal, Vegetable, Miracle,* Barbara Kingsolver (2007) documents her own family's experience of growing and buying local, seasonal organic foods. Her book and website have a wealth of farming and food related links (mostly USA based, but they will have links to organizations across the world).

(www.animalvegetablemiracle.com; www.permaculture.com) [7]

Farmer Managed Natural Regeneration

Tony Rinaudo, of World Vision, Australia, sent me an interesting story of an African reforestation programme. I will go into this in some detail, as it provides important lessons for sustainable development practice, and reforestation.

Rinaudo's article on Farmer Managed Natural Regeneration (Leisa, 2007) states: 'Conventional methods of reforestation in Africa have often failed. Even community-based projects with individual or community nurseries struggle to keep up the momentum once project funding ends. The obstacles working against reforestation are enormous. But a new method of reforestation called Farmer Managed Natural Regeneration (FMNR) could change this situation. It has already done so in the Republic of Niger, one of the world's poorest nations, where more than three million hectares have been re-vegetated using this method.'

Similar programmes are being initiated in other African countries.

The Niger was an area hit hard by desertification, as people chopped down trees and vegetation for firewood, and because they believed that trees competed with their crops. Trees provide us with many benefits in addition to reducing CO_2. For the farmers, the ecosystem of the trees provided: natural predators – which reduced the insects that ate the crops; fertilizer from the creatures that sheltered in their shade; as well as protection from extreme heat and wind. With the trees gone, crops were devastated by drought and insects, and famine spread across the land. Programmes across Africa attempted deforestation programmes that involved growing the trees from seed and planting them in the damaged areas. However, despite investing millions of dollars and

Inspiring Examples: Sustainable Living

thousands of hours labour, there was little overall impact. The conditions were harsh for the trees, not only because of the elements and the animals, but because people continued to cut them down.

The FMNR programme had two crucial allies in the success of their project: the one was the earth and the other was the local farmers. By observing the shoots that came out of some of the felled stumps, Rinaudo became aware of the 'underground forest' that could be harnessed. Careful pruning could support the trees to regenerate. These ancient methods (coppicing and pollarding) were taught to small-scale local farmers; and accompanied by support and education programmes, as well as laws that both protected trees and allowed farmers to harvest them in a sustainable fashion.

'The benefits of FMNR quickly became apparent and farmers themselves became the chief proponents as they talked amongst themselves. FMNR can directly alleviate poverty, rural migration, chronic hunger and even famine in a wide range of rural settings. FMNR contributes to stress reduction and nutrition of livestock, and contributes directly and indirectly to both the availability and quality of fodder... The environment in general benefits as bio-diversity increases and natural processes begin to function again.'

Malatin André, a Chadian farmer practicing FMNR for just two years reported: 'Food production has doubled and many people who were laughing at us, have also adopted the techniques for soil regeneration. As a result, there is always good production, the soil is protected from erosion and heat, and women can still get firewood' (Rinaudo, 2007).

(www.leisa.info) [8]

The FMNR programme is an example of how appropriate sustainable practices offer climate change mitigation as well as adaptation benefits.

It reminds us of a truth that is at the core of most successful sustainable living practices (including those discussed in this chapter): the solutions to a problem are usually right in front of us – in the Earth and in the people. We should work closely with these resources, rather than imposing programmes that go against the grain.

How Can You Make a Difference?

Educate, organise and act

Educate yourself and others. Discuss climate change in your own sector, organization, social circle or family. Locate an organization that you like, and join or support them. Research, plan and take action. Target the high carbon-dioxide producers and persuade them to make changes.

Pressurise government to regulate energy production and consumption

Push government to take the necessary action (see Climate Change Petition). Support organizations that are doing this.

Apply pressure to industry

Consumers, workers or businesses can encourage or pressurize industry to use renewable energy, and to make products and adopt practices that are energy-efficient. If possible, do this through consumer, trade union or other organizations.

Reduce your CO_2 footprint

Some ideas:
- Audit and reduce your carbon footprint in all aspects of your life.
- Minimize car and aeroplane travel: walk, cycle, and use public transport.
- Switch to a low-emission vehicle. You can adapt your current vehicle to hybrid or electric, and use technology that reduces your emissions (such as an HHO-electrolyzer or Naf-Tech).

110

How Can You Make a Difference?

- Be energy-efficient and use efficient and renewable technology.
- Turn off lights and all appliances (including your geyser) when not in use.
- Install a solar water heater and solar PV panels (or other forms of renewable energy) in your home.
- Cook with a hotbox or solar oven.
- Reduce electricity use in peak times.
- Eat organic and free-range, locally-produced, seasonal, unprocessed food.
- Reduce unnecessary purchases, and buy products with low carbon footprints.
- Compost and recycle waste, and conserve water.
- Have fewer children (especially if you are rich).
- For more ideas, see: www.onehundredmonths.org; *The Climate Diet: How You Can Cut Carbon, Cut Costs and Save the Planet* (Harrington, 2008, www.climatediet. com) and *Bending the Curve* [Zipplies (ed.), 2008, www.bendingthecurve.co.za]. The latter book outlines what you can do depending on your occupation (e.g. farmer, educator, business, government etc.).

Climate Change Petition

Signing a petition is one simple way of adding to the pressure on government and business to take action. Yes, it is just a piece of paper, and it should be accompanied by other actions, but it helps us to clarify what needs to be done. You may want to use or adapt the petition I have included here, adding specific demands relevant to your country and conditions. Collect thousands of signatures or just add your own, and send it to your government.

You could also set up an email or website version of your petition. Send or deliver it to the appropriate people and departments. The best people may be the President/ Prime Minister and the Minister of Environment; but you will need to research this in your own country.

CLIMATE CHANGE PETITION TO GOVERNMENT

There is already a 20 to 30% chance that we will exceed a 2° C rise in global warming. We need to take actions that will keep this risk as low as possible.

We call on you to urgently:

- Commit to reducing greenhouse gas (GHG) emissions, at the rate required by science, to avoid the likelihood of a 2° C rise in global warming, i.e. bring the world GHG levels down to below 400ppm CO_2 equivalent, and the carbon dioxide concentration down to below 350ppm CO_2.

- Cut emissions by means of the environmentally friendly use of: energy efficiency; energy-efficient technology; and renewable energy.

- Switch over to 100% renewable energy use, by 2018.

- Introduce feedback tariffs, allowing households and small companies to sell their renewable energy to the electricity grid.

- Put an immediate moratorium on: the mining of fossil fuels (coal, oil and nuclear); opening new fossil-driven power plants; the production of new cars using fossil fuels; and deforestation of indigenous forests.

- Only allow forest harvesting and agricultural practices that are environmentally friendly and produce minimal GHGs.

- Set up urgent, binding international agreements with stringent short, medium and long-term emission-reduction targets for all countries.

- Adopt climate change adaptation measures that are environmentally friendly and support mitigation practices.

Afterword by a Fire Goat

There is still a lot of work ahead for the fire dogs of climate change, the guardians and the watchdogs of the Earth – but we have good reason to feel joy in our hearts, and a wag in our tails. The barking was heard, and the battle is on.

When I started writing these stories in 2006, although scientists and environmentalists were aware of climate change, it was not a major issue for the general public or national governments. But, as predicted by astrologers, the Year of the Fire Dog (2006 – 2007) was one in which there was much 'fierce barking'. There were indeed debates about how to save the planet, and governments were pushed to change their policies. New reports were released by the Intergovernmental Panel on Climate Change, books and movies were produced about the dangers of global warming, and packs of fire dogs growled and barked in warning and protest. Climate change became one of the hottest issues addressed by the media, governments, businesses and civil society.

2007 was the Chinese Year of the Fire Pig (representing the wild boar, rather than the capitalist pig). This year was predicted to be one of clashes of value systems and changes of attitudes, and indeed it was.

2008 was the Year of the diligent, inventive Earth Rat. (The second last Year of the Earth Rat was 1888, in which Nicolas Tesla invented the electric motor.) Rats are excellent organizers and sharp on issues of food and survival. Protests against government and businesses who are threatening our survival continued, and there was ongoing transformation around climate change issues. Some of the changes appeared to be slow because they took place 'underground' or 'behind the scenes' (where rats love to organize). It was also a year in which communities looked at how to change their lifestyles. [1]

Today, the hard work of the fire dogs is bearing fruit. Many developed countries, and a few developing countries are setting targets for significant CO_2 reductions. China contains over a quarter of the world's population and is one of the biggest, and fastest-growing GHG emitters. Let's hope they support the fire dogs in the battle to cool the heating Earth.

From February 2009, it is the Chinese year of the Earth Ox, representing diligent duty and hard work. Will this give us the strength to tackle the recon-

Afterword by a Fire Goat

struction required of us, and to ensure the December 2009 international agreements go far enough?

This is followed by the Year of the Metal Tiger in 2010. What will we make of this powerful, passionate year? And of the next handful of years to come, that will determine the fate of the next thousands of years that follow?

Only after writing this book, did I look up my personal Chinese Year sign, and discover that I am not a fire dog, but a fire goat! *Baaa aaaaa.* At first I found this a bit embarrassing, and then, as I read more, I came to understand its truth. Although I've often run with the fire dogs, I have found that I burn out fast. I am happier meandering over rocky mountains. The 'goat' represents the 'creative artist'. Through this book, I came to find my best role (my 'dance with fire') to be writing stories for dogs, rather than running with the pack.

I hope *The Fire Dogs of Climate Change* reminds you of your own fire, and that your heart leads your feet on a good walk (or run) across the Earth. This Gaia, of which we are an integral part.

Educator's Guide

Reading a book is itself an educational process. Learning takes place through feeling, thinking and doing, and *The Fire Dogs of Climate Change* aims to inspire all three activities. Just sharing this book with others will facilitate climate change education. However, some of you may be involved more directly in educational work, and wish to use *Fire Dogs* as a teaching tool.

Fire Dogs is aimed at adults and older youth (age 14 upwards). The themes addressed in the book relate to environmental issues – climate change in particular. They provide information on problems as well as solutions. They encourage readers to take political and personal action. They also engage the reader on a psychological and spiritual level, encouraging them to open their hearts. The book can be used to explore all of these themes, and to develop language, communication and writing skills.

Here are some recommendations. Adapt them to suit your audience, setting, purpose and curriculum requirements.

- Respect your own knowledge and the knowledge of others.
- Be guided by the clarity of your heart.
- When you have read a story (on your own, or out loud) allow for a few minutes break. Silence, music, drawing, or gazing into a fire are good. Stories are similar to life experience – they need time to sink in.
- The information and action sections of the book, however, should be followed by questions, discussions and debates. The fact sheets contain technical information that people need to understand clearly.
- The stories are useful prompts for creative writing, and for discussion of personal perspectives.

Educator's Guide

- The inspiring examples are good ways of triggering further research and action.
- Questions are a useful way of getting people to engage with the issues. Ideally the learner and the educator should come up with questions or topics of their own, after reading each chapter. It is often more effective to allow people to discuss subjects in pairs or small groups, before sharing with a bigger group.
- You can also use a range of *popular education* methodologies to explore a theme further. (Popular education tools include story telling, pictures, plays, debates, dramatizing radio and TV interviews, etc.)
- After every story you can ask, 'What was this about? What did it make you feel or think? Can you think of a similar personal story?'
- After each fact sheet, you can ask, 'Are there sections you don't understand, disagree with or would like to discuss further. Are there other facts that you need to research?'
- After every 'Inspiring Examples' chapter, you can ask, 'Do you have other examples to share? Is there something here that you would like to do?'
- See 'Educational Resources' listed at the end of this book.

Here are some more specific ideas of questions and activities you could ask or do, after reading each chapter of the book. I structure them as if you are working through the *Fire Dogs* book chronologically.

Introduction

After you have read 'About the Author' and 'Introduction', ask, 'What is climate change? What do you think we need to do about it?'

Write down answers individually (and anonymously) for the educator to read. Then allow for a broad discussion. Use this to assess the knowledge and opinions amongst your group. This will shape what you focus on in future education sessions.

Foreword by Gaia

Play a piece of music that you think resonates with this prose. Write a song, beginning with the words, 'My name is... ' Don't pressurize people to read their songs out loud, but encourage anyone who would like to take it further, to set it to music.

Write one page on: 'Is the Earth a living being?'

The Fire Dogs of Climate Change

Reading Stories to Fire Dogs

Write for three minutes without stopping (*free-writing*) on the subject of 'When I grow up...' If you cannot think of what to write, you must not put down your pen, but just rewrite the subject again and again, until something comes up.

Draw a picture in response to the question: 'Which part of your self or your society do you listen to in making the decision (about when you grow up)?'

Tails of Integrity

Discuss in pairs, then report back to bigger group: 'What is the nature of your relationship to the Earth? What does integrity mean for you? How do you feel about the possible death of the Earth? What are the losses you have had to face in your life? How have you dealt with them? What works well for you?'

Fact Sheet 1: The Problem

Discuss in the large group: 'Are there still people who are ignorant about or deny climate change? If so, why is this? Who stands to benefit from saying climate change is not caused by fossil fuels?'

'What is the primary cause of global warming? Who or what has caused it? Try to be specific in relation to sectors of society (rich/poor/industry/government etc.).'

Prepare a debate in which different 'experts' on the panel disagree about the causes of climate change. For example, 'Professor 1' says 'overpopulation' is the cause of high energy consumption, 'Prof 2' says 'industry' and 'Prof 3' says 'the middle classes'. Have a chair for the session. Allow each panelist a two-minute presentation; followed by questions and comments from each other, and then questions and comments from the floor.

Riding the Moon

Do free-writing for five minutes on: 'What makes my heart sing.'
Write a paragraph on: 'Is it important to experience both sorrow and joy? Why?'

Fact Sheet 2: The Solutions

Look at each proposed solution and discuss its implications. Ask, 'Who should implement the solutions?' Allow time for students to research specific topics and prepare in-

Educator's Guide

formed debates on contentious issues (see current newspapers). Examples of debates: 'Renewable energy vs. nuclear power'; or 'Is there such a thing as "clean coal?" '

It is important that both sides prepare well so the debate is gripping and makes the audience think. A good debate should contain reliable facts and address ethical questions.

Running at the Enemy

Discuss in groups of four, then report back to the group: 'Is it important to believe you can make a difference? Why?' Prepare a short play, which illustrates this point.

Choose an environmental problem in your area. Draw a picture that shows people avoiding (or doing nothing about) the problem, then another picture that shows people confronting the problem.

Inspiring Examples: Action

'What political action has taken place in your country? And in your community and workplace? Are there any lessons you could learn from the inspiring examples listed – or that they could learn from you? Are there any actions you would like to initiate or join in with? Where? When? How?'

Design three banners or placards with slogans that you might use for this occasion. Make one of them now.

Dancing with Fire

Discuss: 'What can we do to make good? How can we do this in a way that also brings joy to ourselves?'

Perform a play that shows 'the problem', and how it could continue in the worst-case scenario. Act out the play once, and then introduce a 'facilitator', who invites the audience to play characters that intervene and effect the play at any point. Do this a few times, so you can play with a few different 'solutions'. Perhaps you can bring in the banners from the previous chapter?

Inspiring Examples: Groundbreaking Technology

Discuss: 'Why does some technology dominate the market? Who decides what gets produced and sold? What influence do you have over it? Are there any technologies, listed here or elsewhere that you think should be in frequent use?' Here's the chance for

The Fire Dogs of Climate Change

science students and techno-whizzes to have some fun. Select one or two technologies and research them in depth. Present the information to others. 'How do you assess if the technology is really 'green'? What can you do to ensure green technology is used?'

Inspiring Examples: Sustainable Living

'Do you know of other inspiring examples of sustainable living, closer to you?' Research (if possible, visit) a project in your country, preferably close to you. Present this information to others using pictures and/or sound.

How can *you* make a difference?

Discuss in groups of three, and then report to the bigger group:
- 'What motivates us to act? Is it heart, mind, body, spirit, law or force?' Do scare tactics work, or make us want to deny the evidence?'
- 'What thing, big or small, are you happy to do as an individual? Is there something that you could give up to mitigate climate change? Is there something that you could gain by taking action?' Be specific about your plans: give details and dates.
- 'What thing big or small will you do as a group?' Be specific: provide details and dates.

Discuss what problems you might face in making your plans happen. Suggest some ways of overcoming them.

Set up a system of how to support each other to ensure the actions are done. 'How can you ensure your intentions become reality?' (e.g. charts on the wall, buddy systems, check-up systems, follow up meetings, rewards, announcing your intentions to a big group of people, etc.).

Write a futuristic story in which people are living in a way that is much kinder to the Earth. Go into some detail about some of the technology or sustainable living practices. Have a competition with a prize for the best story.

Petition

'What do you think of the contents of this petition? To whom would you like to send it? If you were the person getting the petition, what would make you take it to heart or discard it? Are there any changes you would like to make? Will you collect signatures or just send the petition from your group? If you are collecting signatures, how many are you aiming for, and how will you collect them?'

Educator's Guide

Afterword by a Fire Goat

'What do you think has been happening over the last few years in relation to climate change. What do you hope will happen over the next couple of years? What role would you like to play?'

Educator's guide

'Can you educate others about climate change? What resources can you use to help you?' Draw up your own design for a one-hour climate change education session with students who are a bit younger than you. Make it fun.

Prepare a 40-minute power-point presentation to a group of pro-nuclear scientists (a bit older than you) arguing that renewable energy is better than nuclear. Make it fun.

Resources and References

This includes a list of educational resources and programmes.

121

Thanks

Thank you to my wonderful friends, family and colleagues for all your support! You provided fuel and sparks for *The Fire Dogs of Climate Change*, and you fanned the flames to make them strong.

My parents Bosky and Paul Andrew were my greatest fans; Bowen Boshier, my main man. Scores of fantastic people gave me warm and wise feedback, including Eureta Rosenberg, Peter van Straten, Ally Ashwell, Kate and Geoff Davies, Steve Thorne, Glen Ashton, John Yeld, Arona Dison, Hettie and Betty Gets, Joan van Gogh, Jim Taylor, Clare Peddie, Al Gore, the students on the Rhodes/Gold Fields Environmental Educators' course and Sam Humphries. David Parry-Davies was a spark-provider for 'Reading Stories to Fire Dogs', and 'Foreword by Gaia'.

I am grateful to the fire dogs who shared their stories with me: Desmond D'sa of the South Durban Community Environmental Alliance; Bowen Boshier; the Gwich'in Steering Committee of Alaska and the Sea Shepherd crew on the *Farley Mowat*.

Fire dog individuals and organizations from around the world responded to my requests for examples of political action, technology, and sustainable living. I have featured most of the contributions (sadly, I did not have space for all) in the 'Inspiring Examples' chapters. I have drawn on a number of websites that provided me with examples and quotes (acknowledged in these same chapters).

Members of Earthlife Africa and the South African Climate Action Network provided factual input, and helped me preen the scruffy dogs before they went out into the world. I am particularly grateful to Rob Zipplies, Barry Kaplan, Harald Winkler, Wally Menne, Katherine Bunney, Cabral Wicht, Joy Robinson, Jonelle Naudé, Maya Aberman and Bobby Peek.

Thank you to my Findhorn editor, Michael Hawkins, who did the final grooming of *The Fire Dogs of Climate Change* with gentleness and care, until their coats gleamed.

I had wonderful research assistants, and expert guidance. Alistair Gets is a sustainable energy engineer who gathered data and checked on all of the 'groundbreaking technologies.' Techno-wizard Bowen Boshier brewed up a whole cauldron of stuff for me; and journalist Leslie Berliant (Celsias) provided me with a

Thanks

passel of her favourites. *Popular Mechanics South Africa* and Peswiki were great sources of innovative technologies.

David Simpson, Daniel Wahl, Gavin Lawson, Jonathan Dawson, James Blignaut, Alexis Ringwald, Meredith Niles, Colleen Kredell, Brett Simpson and Helmut Hirsch went out of their way to provide good ideas.

Ace educators, Sue Brundrit and Patrick Dowling gave me comments on the 'Educators Guide'. Environmental consultant, Catherine Fedorsky, provided feedback on 'Inspiring Examples: Sustainable Living.' UK activists, Rose Rickford and Alana Dave shared handfuls of action examples. Alana gave input and feedback on the whole book, and was my number one hand-holder and adviser.

I am grateful to Jenny Wheeldon, Mark McClellan, Meg Jordi, Maire Fisher, Helen Burton, Vincent Silimela, Tessa Oliver, Crecilda van den Berg, Sharon Wallbran, Robin Stuart-Clark, and my literary agent, Ron Erwin, for assisting with the production and promotion of the earlier version of this book (*Stories for Fire Dogs*).

Acknowledgements are due to Hot Press, who first published these stories and fact sheets in *Stories for Fire Dogs (Opening our hearts to the Earth)*. Also to Pan Macmillan (1987) for permission to use quotes from the *Gaia Atlas of Planet Management*. Mapping Worlds and Palgrave Macmillan gave permission to reproduce: 'Mapping the global variation in CO_2 emissions' and 'Rich countries – deep carbon footprints' (United Nations Development Programme, *Human Development Report 2007/2008: Fighting Climate Change: Human Solidarity in a Divided World*, Palgrave Macmillan, 2007). World Resources Institute supplied the marvellous 'spaghetti' chart, '*Navigating the Numbers*'.

The following people and organizsations allowed me to quote substantial extracts from their documents and websites: Al Gore (www.wecansolveit.org); Wisions (www.wisions.net); Anna Rose (Its Getting Hot in Here); Leslie Berliant (Celsias); *Popular Mechanics South Africa*, Peswiki, Greenpeace/ EREC (2007); Tony Rinaudo and LEISA magazine. Al Gore's quotes are excerpted from his speech on Renewable Energy, July 17, 2008 at The Daughters of the American Revolution's Constitution Hall. Copyright © 2008 by Al Gore, and reprinted with permission of The Wylie Agency, Inc.

I am also grateful to the Global Greengrants Fund and Earthlife Africa Cape Town for contributing funding towards the first (Hot Press) edition of this book.

Thanks to Sabine Weeke, Carol Shaw, Thierry Bogliolo and the whole team at Findhorn Press for helping *Fire Dogs* to gallop all over the globe.

Please note that my thanks do not hold any of the above people or organizations liable for the contents of this book: all opinions and errors are mine.

References and Resources

Here are further details for the references I have drawn on, as well as resources I recommend.

For direct online access to the websites listed here (and elsewhere in this book) go to http://firedogs.findhornpress.com (These links will be updated regularly.)

References

Anderson, J. (editor). *Towards Gondwana Alive: Promoting biodiversity and stemming the Sixth Extinction*, Cape Town: Gondwana Alive, 1999.

Environmental and Energy Study Institute, *Fact sheet: Jobs from Renewable Energy & Energy Efficiency*, Washington: EESI, 2007 (fbeck@eesi.org).

Flannery, T. *The Weather Makers*, London: Penguin, 2006.

Gore, A. *'A Generational Challenge to Repower America'*, Speech at the Daughters of the American Revolution's Constitution Hall, USA, July 17 2008 (www.wecansolveit.org).

Gore, A. *An Inconvenient Truth*, USA: Rodale Books, 2006.

Greenpeace/EREC (European Renewable Energy Council), *Energy [R]evolution: A Sustainable World Energy Outlook*, EREC, 2007 (www.energyblueprint.info).

Hare, B. and Meinshausen, M. *How much warming are we committed to and how much can be avoided?* PIK report 93, Potsdam Institute for Climate Impact Research, 2004.

Harrington, J. *The Climate Diet: How You Can Cut Carbon, Cuts Costs and Save the Planet*, Earthscan, 2008 (www.climatediet.com).

Kingsolver, B. *Animal, Vegetable, Miracle: Our year of seasonal eating*, London: Faber and Faber, 2007.

References and Resources

Lang, S. 'Cornell ecologists study finds that producing ethanol and biodiesel from corn and other crops is not worth the energy', *Cornell University News Service*, July 2005.

Leaky, R and Lewin, R. *The Sixth Extinction: Biodiversity and its survival*, Canada: Weidenfield and Nicolson, 1995.

Lohmann, L. 'Carbon trading: a critical conversation on climate change, privatisation and power', *Development and Dialogue,* September 2006 (www.dhf.uu.se/publications.html).

Lovelock, J. *The Revenge of Gaia*, London: Allen Lane, Penguin, 2006.

Millennium Ecosystem Assessment Report, Island Press, 2005.

Monbiot, G. *Heat: How to stop the planet burning*, London: Penguin, 2006.

Monbiot, G. 'Apart from used chip fat, there is no such thing as a sustainable biofuels,' *The Guardian*, 12 February 2008 (www.guardian.co.uk/commentisfree/2008/feb/12/biofuels.energy).

Myers, N. (editor). *Gaia Atlas of Planet Management*, London: Pan Books, 1985.

Oxfam International, 'Another inconvenient truth: How biofuel policies are deepening poverty and accelerating climate change', June 2008 (www.oxfam.org.uk/resources/policy/climate_change/bp114_inconvenient_truth.html).

Rinaudo, T. 'Farmer Managed Natural Regeneration,' *Leisa Magazine*, 23.2, June 2007 (http://ileia.leisa.info/index.php?url=article-details.tpl&p[_id]=113390).

Stern, N. *The Economics of Climate Change: The Stern Review*, Cambridge University Press, 2006.

United Nations Development Programme, *Human Development Report 2007/2008: Fighting Climate Change: Human Solidarity in a Divided World*, Palgrave Macmillan, 2007 (http://hdr.undp.org/en/).

United Nations, *Intergovernmental Panel on Climate Change Fourth Assessment Report* (www.ipcc.ch).

Zipplies, R. (editor). *Bending the Curve: Your guide to tackling Climate Change in South Africa*, Cape Town: Africa Geographic, 2008 (www.bendingthecurve.co.za).

Movies

- *An Inconvenient Truth*
- *Who Killed the Electric Car?*
- *The End of Suburbia*
- *The Corporation*
- *The Planet*
- *The 11th Hour: Turning Mankind's Darkest Hour into its Finest*
- *The Story of Stuff*

The Fire Dogs of Climate Change

General information and data

- Intergovernmental Panel on Climate Change: www.ipcc.ch/
- Emissions Database for Global Atmospheric Research: http://www.rivm.nl/en/milieu/
- Mapping Worlds: www.mappingworlds.org
- World Resources Institute: www.wri.org
- www.worldwatch.org
- www.theglobaleducationproject.org
- Climate Action Network: www.climatenetwork.org
- Inconvenient Truth: www.climatecrisis.net
- Tyndall centre for climate change research: www.tyndall.ac.uk
- Transnational Institute: www.tni.org
- Pew Center on Global Climate Change: www.pewclimate.org
- International Energy Agency (IEA): www.iea.org/textbase/pm/index_clim.html
- Climate list: www.climate-l.org
- Climate Change and Energy Group at IISD (International Institute for Sustainable Development): www.iisd.org/climate
- United Nations Environment Program: www.unep.org
- United Nations Framework Convention on Climate Change: http://unfccc.int/
  National Communications for Annex 1 countries: http://unfccc.int/national_reports/annex_i_natcom/submitted_natcom/items/3625.php
  National Communications for Non Annex 1 countries:
  http://unfccc.int/national_reports/non-annex_i_natcom /items/2979.php
- International Institute for Applied Systems Analysis: www.iiasa.ac.at
- International Energy Agency, Greenhouse Gas Programme: www.ieagreen.org.uk
- Climate Analysis Indicators Tool (CAIT), World Resources Institute: http://cait.wri.org

Country specific government climate change data:

- Australian Department of Climate Change: www.greenhouse.gov.au/
- Environment Canada Climate Change Page: www.ec.gc.ca/climate/home-e.html
- China's National Climate Change Programme: www.pewclimate.org/docUploads/ChinaNationalClimateChangeProgramme%20June%2007.pdf
- European Commission Climate Change Page: http://ec.europa.eu/environment/climat/home_en.htm
- Germany, Ministry for the Environment, Nature Conservation and Nuclear Safety:

References and Resources

Integrated Energy and Climate Change Programme: www.bmu.de/english/climate/
downloads/doc/40589.php
- Japan Ministry of Foreign Affairs, Climate Change: www.mofa.go.jp/policy/
environment/warm/cop/index.html
- Mexico, National Strategy on Climate Change, Executive Summary (in English):
www.semarnat.gob.mx/queessemarnat/politica_ambiental/cambioclimatico/
Documents/enac/sintesis/sintesisejecutiva/Executive Summary.pdf
- UK Climate Change Programme 2006: www.defra.gov.uk/environment/
climatechange/uk/ukccp/index.htm

Carbon Trading

- Carbon Positive: www.carbonpositive.net
- www.dhf.uu.se
- Carbon Trade Watch: www.carbontradewatch.org.
See 'The sky is not the limit: The emerging market in greenhouse gases', 2003
and 'The carbon neutral myth – Offset indulgences for your climate sins', 2007
(www.carbontradewatch.org/pubs/index.html).
- www.thecornerhouse.org.uk/pdf/document/CTvsPos.pdf

Business initiatives

- World Business Council for Sustainable Development: www.wbcsd.org
- www.carbontrust.co.uk/default.ct

Nuclear energy

Research:
- www.lebensministerium.at/article/articleview/56678/1/7031
(go to bottom of page, 'Assessment English')
- www.sdc-commission.org.uk
Organizations:
- www.antenna.nl/wise
- www.stopnuclearpower.org
- www.greenpeace.org

The Fire Dogs of Climate Change

Biofuels

- www.biofuelwatch.org.uk
- Citizens United for Renewable Energy & Sustainability: www.cures-network.org
- www.carbontradewatch.org/pubs/Agrofuels.pdf
- World Rainforest Movement: www.wrm.org.uy. See 'Biofuels: a potentially positive solution turned into a serious threat to the South' and 'Brazil: Agrofuels represent a new cycle of devastation of the Amazon and Cerrado regions' WRM Bulletin Issue 116, March 2007.

Organizations

- See organizations listed under 'Inspiring Examples: Action', and 'Inspiring Examples: Sustainable Living'.
- www.wwf.org.uk
- www.climatecommunity.org
- www.1sky.org
- www.earthlife-ct.org.za
- www.bicycology.org.uk
- www.indymedia.org.uk
- www.criticalmass.info
- www.climateimc.org
- www.platformlondon.org
- www.seedsforchange.org.uk
- www.sinkswatch.org
- www.climate-speakers.org.uk
- Project 90 by 2030: www.project90x2030.org.za

Sustainable living and technology

- See websites listed under 'Inspiring Examples: Groundbreaking Technology' and 'Inspiring Examples: Sustainable Living'.
- Rocky Mountain Institute: www.rmi.org
- Sustainable Energy Africa: www.sustainable.org.za
- Sustainable Energy and Climate Change Partnership: www.earthlife.org.za/seccp
- Enviropaedia: www.enviropaedia.com
- Centre for Alternative Technology: www.cat.org.uk
- Zero Carbon Britain: www.zerocarbonbritain.com

References and Resources

- Ecofootprint: www.ecofoot.org
- The Simple Living Network: www.simpleliving.net
- Slow Food: www.slowfood.com

Educational resources

- See 'Fun plays and educational projects' under 'Inspiring Examples: Action'.
- See the educational institutions and networks listed as: 'Institutions sharing ideas and training', under 'Inspiring Examples: Sustainable Living'.
- Education for Sustainable Development: www.unesco.org/education/desd
- United Nations Environment Programme: www.unep.org
- Al Gore's study guide for *An Inconvenient Truth* on: www.climatecrisis.net
- See 'You the Educator' by Tess Fairweather in Zipplies (ed.), 2008. The below educational resources are from her chapter.
- The National Association for Environmental Education (UK): www.naee.org.uk
- Schumacher College: www.schumachercollege.org.uk/learning-resources
- Australian Association for Environmental Education: www.aaee.org.au
- Schools For a Sustainable Future: www.sfsf.com.au
- Department of Environment and Tourism, South Africa: www.environment.gov.za/ClimateChange2005/Resources_schools.htm
- Preschool environmental education: www.wyongsc.nsw.gov.au/environment/preschool_environmental_edu.html
- Centre for Ecoliteracy: www.ecoliteracy.org
- Seed – Transforming Learning through Permaculture: www.seed.org.za
- Earth Education: www.eartheducation.org
- Foundation for Environmental Education: www.fee-international.org

Endnotes

For direct online access to the websites listed here (and elsewhere in this book) go to http://firedogs.findhornpress.com

Reading Stories to Fire Dogs

1 *rooibos tea* – red bush tea (Afrikaans)
2 *goggas* – little creatures or insects (Afrikaans)
3 The friend was David Parry-Davies, editor of the Enviropaedia.
4 This information on Chinese astrology is drawn from the website of Richard Giles (www.astrologycom.com/firedog1.html).
5 I first read this story to environmental educators attending a Rhodes/Gold Fields national workshop on 3 February 2006 – New Year's Eve in the Chinese calendar.

Tails of Integrity

1 *Nozizwe* is my isiXhosa name, meaning 'with (or mother of) nations.'
2 *Kunjani* – How are you? (isiXhosa)
3 *Ndinolusizi, nyani* – I am sorry, truly (isiXhosa).
4 Bushmen – Otherwise known as San or Khoisan: Southern African hunter-gatherer communities, most of which have been destroyed.
5 Myers, N (ed.). *Gaia Atlas of Planet Management*. Pan Books, London, 1985.
6 *sommer* – casually (Afrikaans)
7 *gemsbok* – oryx (Afrikaans): a large wild buck found in Southern Africa

Endnotes

Fact Sheet 1: The Problem – Environmental Destruction and Climate Change

1 See also 'Big Picture', 2006: www.theglobaleducationproject.org
2 See www.foei.org for more examples.
3 www.wri.org/publication/navigating-the-numbers

Riding the Moon

1 *Fynbos* literally means 'fine bush' (Afrikaans). It is part of the highly diverse and endangered floral kingdom, found predominantly in the Western Cape, South Africa. This is an international biodiversity hotspot, and includes the 'protea' flowers.
2 This was reported in the *Sunday Independent*, June 2006.
3 *liedjies* – little songs (Afrikaans)

Fact Sheet 2: The Solutions – Energy Efficiency and Renewable Energy

1 For a detailed example, see 'Farmer managed natural regeneration' under 'Inspiring Examples: Sustainable Living'.
2 See table 5.1, Ch. 5, IPCC AR4 report, www.ipcc.ch
3 See table 13.7, Ch. 13, IPCC AR4 report, www.ipcc.ch
4 Al Gore, 2008, www.wecansolveit.org
5 This report can be accessed at: www.lebensministerium.at/article/ articleview/56678/1/7031 (go to bottom of page, 'Assessment English'). See also nuclear websites listed under 'References and Resources'.
6 www.guardian.co.uk/environment
7 See www.cures-network.org and 'Inspiring Examples: Groundbreaking Technology' in this book.
8 See 'False Hope – Why carbon capture and storage won't save the climate': http://www.Greanpeace.org/international/press/reports/false-hope
9 See www.thecornerhouse.org.uk/subject/climate
10 www.thecornerhouse.org.uk/pdf/document/CTvsPos.pdf
11 See, for example, 'My other car is a bright green city,' by Alex Steffen, www.worldchanging.com/archives//007016.html
12 www.bp.com/sectiongenericarticle.do?categoryId=9023767&contentId=7044196

The Fire Dogs of Climate Change

Running at the Enemy

1   *vlei* – wetland (Afrikaans)
2   *bakkie* – vehicle with canopy (Afrikaans)
3   Arundhati Roy was quoted on the Earthfirst! website in 2006 (www.earthfirst.org).

Inspiring Examples: Action

1   See Al Gore's full speech: www.wecansolveit.org/content/pages/304/ and US NGO responses to it: http://usclimatenetwork.org/media/responses-to-gore-speech
2   See www.unep.org/climateneutral and www.unep.org/themes/climatechange
3   See www.unep.org/wed/2008/english/Around_the_World/Africa
4   See www.unions.org/pdf/ohsewpNews#1Pearson_P_13Bc.EN.pdf
5   See, for example, the British Trade Union Council's website on the subject: www.tuc.org.uk/theme/index.cfm?theme=sustainableworkplace
6   See www.tradeunionsdunit.org/profiles/profiles.php
7   To find out more about trade union thinking and actions on climate change go to: www.unep.org/labour_environment/PDFs/TOT-Trade-union-action-on-climate-change.ppt. Also look up 'sustainable development' or 'climate change' on these websites: Trade Union Advisory Committee (www.tuac.org); and International Trade Union Confederation (www.ituc-csi.org).
8   See www.biofuelwatch.org.uk and www.networkforclimateaction.org.uk
9   See also Caricom Climate Change Centre on www.caribbeanclimate.bz/news.php
10 See 'Greenpeace victories', www.greenpeace.org/international
11 See also www.biofuelwatch.org.uk/declarations.php and www.carbontradewatch.org/pubs/Agrofuels.pdf
12 See 'Greenpeace victories', www.greenpeace.org/international
13 See www.save-our-world.net
14 See www.coinet.org.uk; www.unep.org/Tunza/; www.unep.org/publications/search/pub_details_s.asp?ID=3988; www.youthxchange.net/main/home.asp and www.capefarewellcanada.ca
15 Go to www.iisd.ca/email/subscribe.htm and www.carbonpositive.net/viewarticle.aspx?articleID=
16 See www.iea.org/textbase/pm/index_clim.html
17 www.thecornerhouse.org.uk/subject/climate; www.thecornerhouse.org.uk/pdf/document/CTvsPos.pdf

Endnotes

Dancing with Fire

1 A *sangoma* is an African shaman. This sangoma was Mike Skorpen, who was trained in the Congolese tradition.

2 This year of fire dogs or gods runs from the beginning of February 2006 to the beginning of February 2007.

3 At the end of this year of the fire dog the United States Senate issued a bill intended to permanently protect the Arctic National Wildlife Refuge from drilling. They are expecting resistance from the Republicans, so the fire dogs will keep fighting to protect this wild place of ice (see www.gwichinsteeringcommittee.org and www.savearcticrefuge.org).

4 I have not verified the authenticity of this claim. There are a number of inventions in a similar vein that have been authenticated (see 'Inspiring Examples: Groundbreaking Technology').

5 The inventor requested that I do not use his real name at this stage.

Inspiring Examples: Groundbreaking Technology

1 www.zerocarbonbritain.com

2 www.celsias.com/2008/06/10/waterless-washing-machine/

3 See www.belugagroup.com/News.345.0.html?&cHash=8d7c34a750&tx_ttnews%5Btt_news%5D=505 and www.enn.com/top_stories/article/26772

4 See www.treehugger.com/files/2007/10/turtle_airships.php

5 See also: http://en.wikipedia.org/wiki/Tesla_Roadster

6 *Popular Mechanics SA*, September 2008

7 See also http://nycewheels.com/dahon-mup8-bionx-electric-bike.html; www.flexible-energy.co.za and www.giant-bicycles.com

8 See www.popularmechanics.com/technology/industry/1287316.html

9 See also www.turbotecproducts.com; www.bsrsolar.com; www.sigma-el.com; www.solo-germany.com; www.freenergynews.com and 'Concentrated Solar Power', in this chapter.

10 See also www1.eere.energy.gov/femp/pdfs/bamf_wastewater.pdf; www.mda.state.mn.us/renewable/waste/default.htmenewableenergyworld.com/rea/new; www.renewableenergyworld.com/rea/news/story?id=49123 and 51913 and www.ruralcostarica.com/biogas.html

11 See www.enn.com/top_stories/article/24429

12 www.celsias.com/2008/06/11/green-crude/

13 See www.wwindea.org/home/index.php and

www.bp.com/sectiongenericarticle.do?categoryId=9023767&contentId=7044196

14 See also: www.renewabledevices.com/swift/index.htm and www.bwea.com/
small/index.html

15 See also www.rexresearch.com/kelley/kelley.htm

16 http://us.mti.gov.bw/index.php?option=com_content&task=view&id=548&Itemid
=2&lang=en

17 See http://pesn.com/2005/08/11/9600147_Edison_Stirling_largest_solar

18 www.renewableenergyworld.com/rea/news/reworld/story?id=52024

19 www.solfocus.com/product.php?pid=4

20 www.renewableenergyworld.com/rea/news/story?id=52531

21 www.englishdaily.com.cn/90001/90781/90879/6420910.pdf

22 See www.popularmechanics.com/geothermal

23 See www.science.edu/TechoftheYear/TechoftheYear.htm

24 www.environment.co.za/topic.asp?TOPIC_ID=1315

25 See www.greenboatbateauvert.com/tech-blocks/batteries/ultra-capacitators/
eestor; www.A123systems.com; www.altairnano.com and
www.technologyreveiw.com

26 www.eere.energy.gov/de/compressed_air.html

27 www.economist.com/science/displaystory.cfm?story_id=9539806

28 www.renewableenergyworld.com/rea/news/story?id=52873

29 See www.carlist.com/autonews/2004/toyota_fchv.html and www.toyota.co.jp/en/
tech/environment/fchv/fchv08.html

30 See also Wikipedia (zinc-air battery) and www.arotech.com

31 See www.celsias.com/2008/01/10/are-hydrogen-fuel-cells-the-answer-to-our-
municipal-energy-needs and www.novomer.com

32 See www.3dsolids.com

33 See www.treehugger.com/files/2008/07/wave-power-anaconda-rubber-snake-
alternative-energy.php and www.dexawave.com

34 See www.auto-eng.co.za

35 See www.popularmechanics.com/drivegreen

Inspiring Examples: Sustainable Living

1 www.newstatesman.com/blogs/life-at-findhorn

2 www.gcp-urcm.org/Category/UrbanCarbonManagement

3 For more inspiring applications of clean energy, also look at the 'world clean
energy awards' and the projects funded by WISIONS:
www.cleanenergyawards.com and www.wisions.net

Endnotes

4   Also see the Kenya Biogas project (www.itpower.co.uk) on www.wisions.net

5   For similar projects, have a look at ENERGIA (www.energia.org), which is a network on gender and sustainable energy.

6   Also check out international cycling/transport websites: www.carfree.com; www.cycling.nl; www.itdp.org (Institute for Transport and Development Planning) and www.velo.info

7   Many of the websites mentioned earlier (see 'Institutions sharing ideas and training') contain information about sustainable agriculture and food production.

8   See http://ileia.leisa.info/index.php?url=article-details.tpl&p[_id]=113390
    For more ideas and practices related to the 'Restoration of Natural Capital' see: www.rncalliance.org and www.leisa.info (Low External Input and Sustainable Agriculture).

Afterword by a Fire Goat

1   This information on Chinese astrology is drawn from the website of Richard Giles (www.astrologycom.com/firedog1.html) and the Wikipedia: 'Chinese Astrology'.

Notes

The Fire Dogs of Climate Change

Notes

The Fire Dogs of Climate Change

Notes

The Fire Dogs of Climate Change

FINDHORN PRESS

Books, Card Sets,
CDs & DVDs
that inspire and uplift

For a complete catalogue,
please contact:

Findhorn Press Ltd
305a The Park, Findhorn
Forres IV36 3TE
Scotland, UK

Telephone
+44-(0)1309-690582
Fax
+44-(0)131-777-2711
eMail
info@findhornpress.com

or consult our catalogue online
(with secure order facility) on
www.findhornpress.com

For information on the Findhorn Foundation:
www.findhorn.org